CHROMATOPHORES
AND
COLOR CHANGE

Organismal Biology Series

HOWARD A. BERN

SERIES EDITOR

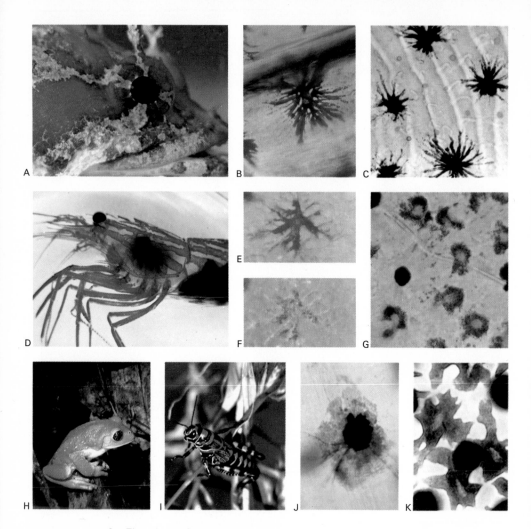

A. The teleost, *Opasanus beta*. (Courtesy of Dr. J. L. Simon and R. R. Lewis III)

B. Erythrophore from the teleost, *Barbus chonchonius*.

C. Melanophores and xanthophores from the teleost, *Fundulus heteroclitus*.

D. The crustacean, *Hippolysmata californica*.

E. Erythrophore of *Hippolysmata* photographed with transmitted light.

F. Same as E, except photographed by both transmitted as well as reflected light.

G. Melanophores and erythrophores of the teleost, *Xiphophorus helleri*. Red pteridine pigments and yellow carotenoid pigments are both present within an individual pigment cell.

H. A Mexican tree frog, *Agalychnis dacnicolor*. (Photographed in the wild by Mr. Thomas Wiewandt)

I. The grasshopper, *Dactylatum variegatum*.

J. Melanophore and iridophore of *Trichogaster trichopterus*.

K. Melanophores and iridophores of the frog, *Rana pipiens*. (Colors in these iridophores are not due to pigments but are structural phenomena.)

CHROMATOPHORES AND COLOR CHANGE

the comparative physiology of animal pigmentation

JOSEPH T. BAGNARA

Department of Biological Sciences
University of Arizona

MAC E. HADLEY

Department of Biological Sciences
University of Arizona

PRENTICE-HALL, INC.
ENGLEWOOD CLIFFS
NEW JERSEY

51262

10 9 8 7 6 5 4 3 2 1

ISBN: 0-13-133603-7

Library of Congress Catalog Number: 72-156

Printed in the United States of America

PRENTICE-HALL INTERNATIONAL, INC., LONDON
PRENTICE-HALL OF AUSTRALIA, PTY. LTD., SYDNEY
PRENTICE-HALL OF CANADA, LTD., TORONTO
PRENTICE-HALL OF INDIA PRIVATE LIMITED, NEW DELHI
PRENTICE-HALL OF JAPAN, INC., TOKYO

to LOU and TRUDY

series statement

organismal biology

Organismal biology, standing as it does with one face toward cellular-molecular biology and the other toward environmental-populational biology, is receiving the full impact of the major advances being made in both these areas. The new information relative to the biology of cells and subcellular systems is clarifying the mechanisms involved in the functioning of the structural complexes within the organism. The new information relative to the ecology of organisms is redefining the importance of the adaptive processes by which organisms make their adjustments to the demands of the environment, both biotic and inanimate.

From within the field of organismal biology itself is a new impetus originating from application of the comparative approach. The comparative approach in biology recognizes the advantages to be derived from the study of similar phenomena in a variety of organisms. However, comparative biology is not pursued for the sake of elucidating similarities and differences alone, but rather in the expectation that we shall learn about the evolutionary events that have led to the structural and functional patterns characteristic of organisms. Only from a deep knowledge of comparative biology can one begin to extract that which is truly general for biological systems. Although the keyword for the comparative biologist is certainly *evolution,* the comparison of analogous adaptive mechanisms that have evolved independently in organisms only distantly related phylogenetically, is also of great value.

This series of books and monographs on comparative biology is intended to provide an up-to-date picture of the biology of organisms, both for the student and the specialist. In presenting a concise and meaningful picture of what is known in the field, the authors also intend to delineate that which is not known—and to lay down the challenge of questions still to be answered by future research and new conceptual approaches.

HOWARD A. BERN

UNIVERSITY OF CALIFORNIA, BERKELEY

foreword

The present volume by Joseph T. Bagnara and Mac E. Hadley deals with the biology of pigmentation—an area of interest to invertebrate and vertebrate zoophysiologists alike. The organism uses its pigment cells—its chromatophores—in a variety of adaptational processes in response to internal and external environmental stimuli. In addition, color patterns may be a consequence of complex developmental events. Color changes and color pattern are organismal phenomena wherein the cellular mechanisms are increasingly better understood. The authors have exploited their own expertise in giving us an up-to-date survey of the phenomena and mechanisms of animal pigmentation. They have done this in a manner calculated to stimulate the interest both of the student and the investigator. Their efforts should provide impetus for further investigations fundamental to understanding this scientifically intriguing and esthetically attractive facet of the living organism.

HOWARD A. BERN
SERIES EDITOR

preface

    This book is an expression of enthusiasm about a subject that has been a source of pleasure to both of us. Although the need for such a book was an important consideration in our undertaking the venture, we suspect that a primary motive was a deep-seated desire to tell biologists and prospective biologists about our research. Hopefully, exuberance for vested areas of interest has not detracted from our principal purpose—to present a balanced review of all facets of animal pigmentation. Fortunately for those interested in pigment cell biology, there has been an excellent source of information in G. H. Parker's book, *Animal Colour Changes.* This work, in part, has served essentially as a bible to many of us, but its value is beginning to diminish because of the availability of much newer biological information that needs to be presented, in digest form, to the biological community. The most important advances have probably been made on both the nature of pigment cells and their hormonal control, and, for this reason, chapters dealing with such subjects are the longest.

    In preparing this presentation, we have deviated from the usual format by using a functional approach rather than the traditional taxonomic one. This approach has presented relatively few problems, for analogies between invertebrate and vertebrate systems are varied and interesting. Difficulties in making comparisons do arise, however, because much more is known about most aspects of vertebrate pigmentation than about that of invertebrates. Moreover, our knowledge about both groups is limited to very few examples. Most of our recent information about vertebrate pigmentation has come from two general groups of workers, each having different points of view. One group, represented by many zoologists working mainly on poikilotherms, has provided most of our knowledge of the control of pigment cells. Such investigators have used color changes as an expression of physiological function and have paid much less attention to the nature of the changes that occurred. Another group of investigators, including many medically oriented scientists, has concentrated on the pigment cell and its relationship to the skin. Naturally, mammals have contributed heavily to their research. It appears, therefore, that although much is known about color change in frogs and about the pigmentation of mice and men, we know different things about each of the two groups. Fortunately, efforts are being made to bridge the communication gap, and the concept of pigment cell biology has now arisen. Hopefully, our presentation adds support to this end by especially emphasizing the physiology of the epidermal melanophore, a chromatophore common to all vertebrate classes.

The contribution of many individuals has made this presentation possible. We are grateful, first of all, to the students and colleagues around us with whom we have shared enthusiasm for research. In this regard, we are especially indebted to Drs. Masataka Obika, Jiro Matsumoto, John D. Taylor, and Mr. Joel Goldman. We are also appreciative of the help provided us by Dr. Giuseppe Prota relative to the chemistry of melanins. To our many scientific colleagues who have kindly provided us with many of the illustrations used in this book, we express our keen appreciation. We are especially grateful to Dr. Richard A. Cloney who provided the electron photomicrograph used on the book jacket.

J.T.B. and M.E.H

TUCSON, ARIZONA

contents

SERIES STATEMENT   vii

FOREWORD   ix

PREFACE   xi

CHAPTER

1   INTRODUCTION   1

CHAPTER

2   THE NATURE OF PIGMENTATION   4

Types of Pigment Cells   6
Special Types of Pigment Cells   25
Cephalopod Chromatophore Organs   27
Physiological Color Changes   28
Morphological Color Changes   31
Dispersion and Aggregation versus Expansion
and Contraction   33
Structural Coloration   34
Assessment of Chromatophore Responses   35
Cellular Associations   37
Patterns   43

xiii

CHAPTER

3

BIOCHEMICAL ASPECTS
OF CHROMATOPHORE PIGMENTS   46

Melanins   46
Ommochromes   50
Purines and Pteridines   51
Carotenoids   54

CHAPTER

4

DEVELOPMENTAL ASPECTS
OF PIGMENTATION   58

Origin of Pigment Cells   58
Migration of Pigment Cells   60
Pattern Formation   62
Determination of Chromatophore Types   65
Hormonal Effects on Chromatophore Development   69

CHAPTER

5

CHROMATOPHORE CONTROL I
PITUITARY ROLE IN THE CONTROL OF VERTEBRATE
CHROMATOPHORE RESPONSES   74

Control of Vertebrate Chromatophore Responses   75
Structure and Bioassay of Chromatophore-
Stimulating Hormones   75
Pars Intermedia (MSH) Regulation of Melanophores   79
Intermedin Regulation of Bright-Colored
Chromatophores   85
Unihumoral and Bihumoral Theories
of Chromatophore Control   88
Regulation of Pars Intermedia Function   89

CHAPTER

6

CHROMATOPHORE CONTROL II
GENERAL ENDOCRINE AND NERVOUS MECHANISMS
OF CHROMATOPHORE CONTROL   98

Pineal Role in the Control of Vertebrate
Chromatophore Responses   98
Direct Neuronal Control of Chromatophores   109
Pharmacological Considerations   112
Role of Steroids in Chromatophore Control   114
Role of the Thyroid in Chromatophore Control   117

CHAPTER

7

CHROMATOPHORE CONTROL III
CONTROL OF INVERTEBRATE COLOR CHANGES   121

Control of Cephalopod Chromatophores   121
Control of Crustacean Chromatophores   123
Control of Color Changes in Insects   127
Color Control in Some Other Invertebrates   129

CHAPTER

8

MECHANISMS OF HORMONE ACTION   132

Structural Requirements for MSH Activity   132
The First Messenger-Second Messenger Hypothesis
of Hormone Action   133
Receptor Mechanisms Regulating Chromatophore
Responses   136
Receptors and Cyclic AMP: Generalized Model
for Chromatophore Control   141
Theories on the Mechanisms of Pigment Granule
Movements   144

CHAPTER

9 **PERSPECTIVES** 160

Nature of Chromatophores   160
Pigment Cell Development   161
Pigment Biochemistry   162
Chromatophore Control   163
Concluding Remarks   166

REFERENCES   167

INDEX   195

CHROMATOPHORES
AND
COLOR CHANGE

CHAPTER

1

introduction

The bright colors of plants and animals have stimulated the esthetic sense of man since the dawn of civilization, so it seems natural that as means became available, the scientifically curious turned their attention to the nature of animal coloration. Aristotle, who dealt with so many areas of science, focused his attention on the remarkable color changes of fishes and lizards. However, true scientific study of animal coloration had to wait until the middle of the nineteenth century when naturalists made the first thorough investigation on color changes of the African chameleon. Many important questions in the physiology of animal coloration were raised and the groundwork was laid for the relevant experiments that followed during the rest of the century.

While the rapid and spectacular color changes of lizards captured the interest of the naturalists, chemists and physicists took up the study of animal colors and made cogent observations on the nature of pigmentation. During this period it was demonstrated that the physical aspects of biological structures are important in the coloration of animals, and the concept of "structural coloration" emerged. Gradually it became apparent that the spectacular iridescent properties of many insects and birds are not due to the presence of blue or green pigments but rather are attributable to interference phenomena. Similarly, iridescent colors can be produced when biological structures such as fibrils or lamellae are arranged in an orderly fashion to serve as a diffraction grating; this is the basis of diffraction colors. A third important phenomenon in "structural coloration" results from the differential scattering of light and is often referred to as *Tyndall scattering*. This phenomenon provides the basis for much of the blue colors that are often seen in association with eyes, feathers, and skins of certain vertebrates.

The chemists were also active during this early period of study and contributed greatly by analyzing extracts of colored tissues. As a result, it was revealed that the lipid-soluble yellow, orange, or red pigments of animals are carotenoid in nature and that the whitish reflecting surfaces of fishes and frogs are composed of guanine deposits. Other studies demonstrated the importance of melanins, flavins, pteridines, and porphyrins in animal pigmentation.

A milestone in our knowledge of pigmentation was the discovery that many of the diverse animal pigments are contained in discrete cells that we now know as chromatophores ("the bearers of pigment"). An important facet of this disclosure was the contention that various pigment cells could change their form to cover a larger or smaller surface area. Thus when chromatophores assume an "expanded" state, the coloration of an animal can be enhanced; and when these cells "contract," the amount of pigmented surface is considerably reduced, thereby resulting in an alteration in color. With the availability of more sophisticated techniques, it has become more obvious that the "expanded" state usually entails migration of pigment granules from the center of the cell to the periphery and that, conversely, "contraction" to the punctate state involves a centripetal movement of these pigments.

Disclosures of the cellular localization of pigments and the movements of pigment granules paved the way for a surge of research. The principal reason was that pigment cells provide both qualitative and quantitative features that can be readily assessed. For example, the investigator occupied with problems of development or differentiation finds an extremely advantageous research material in the many kinds of pigment cells that exist. Many of these cells appear at given stages of development, according to specific pigment patterns. In a similar way, changes that occur to chromatophores in response to physiological stimuli provide a striking subject for the study of regulatory mechanisms. The morphology of pigment cells is also a stimulating subject for investigation; and as a result of advances made in our knowledge of cell biology, the ultrastructure of chromatophores and their pigment-containing organelles has attracted wide attention. These increases in our basic knowledge of pigment cells are of obvious importance to a variety of areas of biological research ranging from genetics to behavior.

As we develop this discussion of animal pigmentation, it will become increasingly obvious that the physiology of pigmentation has attracted the most interest; consequently, a considerable body of knowledge on the control of pigment cells of such widely diverse forms as crustaceans and amphibians now exists. From very early studies it was obvious that the nervous system is an important regulator of color changes, at least in some forms, and we now know that the endocrine system is also a fundamental element in the control of pigment cells of both invertebrates and vertebrates. These hormonally-induced pigmentary events are directed by a variety of glands for a variety of functions. For example, the pituitary, by virtue of its chromatophore–stimulating hormone, can cause darkening of animals during background or temperature adaptation or, by the action of gonadotropic hormones, it may be involved in nuptial coloration. The action of gonadal hormones on nuptial plumage changes is, of course, well known; however, the roles of thyroid or adrenocortical hormones on pigmentation are more obscure.

With these brief introductory comments about what essentially constitutes the physiology of animal pigmentation, it may already be evident why this book is being written. The diverse nature of these problems of color change need to be combined in an integrated fashion that will be of value to expert and student alike. Many specialists in the area of pigmentation are so thoroughly involved in their own specific area of research that they cannot keep abreast of research in areas of pigmentation peripheral to their own. This problem has been compounded by the rapid increase in pigmentary research during recent years. For the student or the reader motivated by academic interest, it is our aim to present the important features of pigmentary research primarily for the sake of knowledge. However, should this information find application in other areas of physiological investigation, our role will be more than justified. In a sense this is one reason for orienting our treatment of pigmentation along comparative lines. To many, the use of the word "comparative" merely signifies that the subject is non-mammalian. Surely this is not our intent; rather, we expect that by making comparisons, the investigator involved in research with one animal group may obtain knowledge about other animal groups that is applicable to his own research. For example, knowledge that frog chromatophores can respond in specific ways to chromatotropic hormones from mammalian sources has been and will continue to be profoundly important to researchers who primarily study pigmentation of either amphibians or mammals. Perhaps one of the more important functions to be served by this book will be an indirect one resulting from the fact that surveying the literature is, in a sense, a kind of introspection. The results of these assessments have dictated the emphasis we have made and have led to the enumeration of research perspectives presented in the final chapter.

the nature of pigmentation

The various color changes that many animals undergo as they respond to their environment result from the movement of pigment granules within cells that have been referred to as chromatophores for over 150 years. Although this statement appears simple and straightforward, in reality it reveals several points of terminological confusion. First of all, while most authors agree about the mobility of pigment granules in some pigment cells, others argue that this may not be a universal mechanism (1). Moreover, pigments are not always in granular form. Use of the term *chromatophore* is the most controversial point of all, for many investigators prefer to use the suffix *-cyte* instead of *-phore*. It is appropriate, therefore, that at the outset an attempt be made to clarify these problems through the establishment of simple terminology that should be useful to the majority of readers. It is not our purpose to champion any specific terms; although this can be done on logical grounds, it seems foolhardy and impractical to attempt to dislodge terminology that has become firmly established. Accordingly, even though we use and recommend various terms, we shall make a point during the course of this presentation of indicating important synonyms. It is most important in the long run to eliminate confusion so that the students of pigmentation may understand one another. Much of the confusion regarding pigment cell terminology is due to the fact that pigment cells are often highly variable in their composition and appearance. In order to facilitate an understanding of the terms we favor, the general characteristics of animal pigment cells have been grouped together and are presented in Table 2–1.

One of the first important policies in establishing a chromatophore terminology is simply to designate pigment cells according to their appearance (2). Thus a black pigment cell should be called black, just as a

T A B L E   2 - 1

Chromatophore terminology

Chromato-phore	Organelle	Pigment	Color	Source
Melanophore	Melanosome	Melanins	Yellow, red, brown, black	Invertebrates and vertebrates
	Granular organelles	Ommo-chromes	Yellow, red, brown, black	Invertebrates
Epidermal melanophore (cyte)	Melanosome	Melanins	Yellow, red, brown, black	Vertebrates, especially homeotherms
Iridophore (Leucophore)	Reflecting platelet	Guanine, adenine, hypoxanthine, uric acid		Vertebrates, especially poikilotherms
	Reflecting platelet (?)	Purines (?)		Crustaceans
	Lamellar ribbons	Unknown		Cephalopods
Xanthophore	Pterinosome	Pteridines, especially sepiapterin	Yellow, orange	Poikilotherms
	Carotenoid vesicles	Carotenoids	Yellow, orange, red	Poikilotherms
Erythrophore	Pterinosome	Pteridines, especially drosopterins	Red, orange	Poikilotherms
	Carotenoid vesicles	Carotenoids	Yellow, orange, red	Poikilotherms

yellow pigment cell should be designated yellow. This step seems more reasonable than to identify pigment cell types according to the chemical composition of their constituent pigment, especially when the same color can be imparted to a cell by completely unrelated substances. An exception to this rule occurs with respect to some epidermal melanophores that contain melanin in the form of yellow or red pigments (phaeomelanins). Since these cells are obviously closely related to the more common epidermal melanophores, they have never been given a separate designation and it would be confusing to do so now.

A second policy that we embrace is an arbitrary one which will not be favored by all readers. We propose that the suffix -*phore* be used preferentially. In this system black pigment cells are designated melanophores (*melanos*: black; *phore*: the bearer of) instead of melanocytes. Surely melanophores are melanocytes, or cells that contain black pigment; however, the former is the term of priority that has remained in continual usage for many years. Moreover, the term melanocyte is not usually applied to pigment cells of invertebrates or to pigment cells of poikilothermic vertebrates. Melanocyte seems to be the term of preference, however, with respect to birds and mammals. This usage has been justified on the basis that these cells are not involved in rapid color changes because their pigment granules do not migrate following hormonal stimulation. This point is now subject to question, for there is evidence that even epidermal melanophores of man contain dispersed pigment after administration of chromatotropic hormones (3). We cannot hope to resolve this problem of terminology to the satisfaction of all concerned; however, it is recommended that the term melanophore be used universally to designate black pigment cells. Pigment cells of other colors pose no problem in this respect because they are usually found only in invertebrates and poikilothermic vertebrates where the suffix -*phore* is generally applied to the satisfaction of all.

TYPES OF PIGMENT CELLS

Melanophores

These black chromatophores are the best known of all pigment cells and are perhaps the most important of the cells active in color changes. Among vertebrates, at least two distinct melanophore types differ markedly from one another, not only by virtue of their location but also by their general appearance and their differential responses to hormones.

Dermal melanophores of vertebrates are involved in rapid color changes and thus are prevalent in poikilotherms. They are found at varying distances beneath the basement lamella and may exist as either relatively flattened cells with radially directed processes (Fig. 2–1A) or they may exhibit a basketlike appearance with processes directed upward from the central plane of the cell (Fig. 2–2). Such melanophores may be very large and are often of the order of several hundred microns in diameter. Generally dermal melanophores appear in far greater numbers in the dorsal integument, which partly explains the fact that the ventral surface of most poikilotherms is light colored. Even on the dorsum, the often-seen patterns of spotting or mottling can be attributed in some degree to localized variations in the numbers of dermal melanophores.

Melanophores are not restricted to the integument alone but are found internally in various organs, and on nerves and blood vessels. These

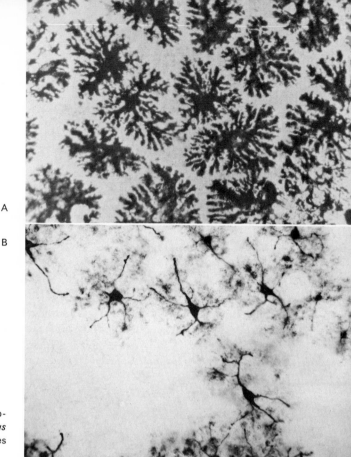

A

B

FIG. 2-1. **A.** Dermal melano-phores of the tadpole, *Xenopus laevis.* **B.** Epidermal melanophores of the frog, *Rana pipiens.*

FIG. 2-2. Cross section of *Rana pipiens* skin. M, melanophore; I, iridophore; X, xanthophore layer. In the epidermis, a single melanophore can be seen in association with epidermal cells containing cytocrine melanin that has been deposited by the melanophore. ×380.

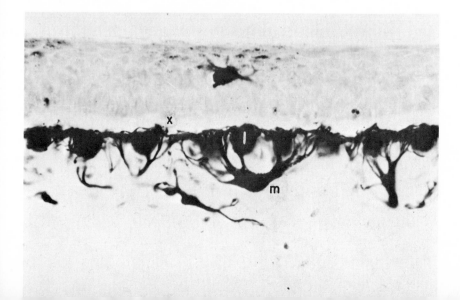

melanophores resemble those of the dermis; and on the basis of their appearance and their sensitivity to hypophysial hormones, it seems probable that these melanophores and dermal melanophores are closely related to one another.

Epidermal melanophores of vertebrates differ markedly from other types of melanophores not only in their shape but also in their sensitivity to certain hormones. This point will be elaborated on later. The appearance of epidermal melanophores is remarkably consistent among all vertebrates (Fig. 2–1B). These cells are generally thin and elongated and are often referred to as "spindle-shaped." Dendritic processes extend outward from the ends of the cell. In amphibian tadpoles there is a definite orientation of epidermal melanophores to form a polygonal network (4). These cells may be extremely long, sometimes reaching 500 μ in length. Under hormonal stimulation their dendritic processes become highly branched, and it almost appears that anastomoses form between the individual epidermal melanophores (Fig. 2–3). Among poikilotherms, these pigment cells are found just above the germinative layer of the epidermis where they may form a relatively uniform layer. Processes or dendrites from these melanophores extend between adjacent epidermal cells and are responsible for the deposition (cytocrine activity) of melanin granules in epidermal cells (Fig. 2–2) (5, 6). This deposition of pigment is relatively slow; thus epidermal melanophores are not of great importance in the rapid color changes of vertebrates. However, prolonged stimulation of epidermal melanophores results in the accumulation of large amounts of cytocrine melanin in the

FIG. 2-3.   Network of epidermal melanophores of a *Rana pipiens* larva.

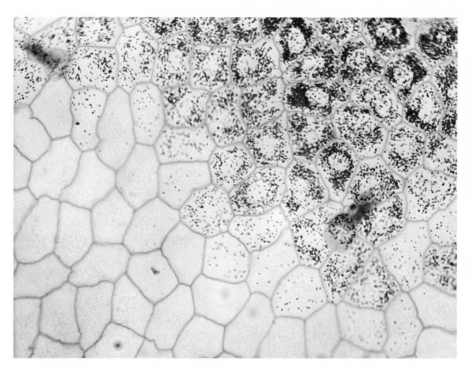

FIG. 2-4. Shed epidermis from the frog, *Rana pipiens,* showing deposits of cytocrine melanin within individual epidermal cells.

epidermis with consequent darkening of the animal (Fig. 2–1B). This pigment is lost slowly as, one by one, successive epidermal layers are shed during molting (Fig. 2–4). In birds and mammals, the counterpart of this cytocrine activity results in the injection of pigment into other epidermal structures, such as the bill, feathers, and hair (7) (Figs. 2–5 and 2–6). The melanin-containing organelle, the melanosome, or melanin granule, may be deposited singly or apparently, in some species, as packets (Fig. 2–6) containing a number of melanosomes. After transfer to epidermal cells, those melanosomes that have been deposited as solitary organelles may, in some races of man, remain free (Negroids) or became grouped into large aggregates of melanosomes (Caucasoids and Mongoloids) (8, 9).

Just as in the dermis, melanophores in the epidermis of poikilotherms are predominantly found in dorsal skin; moreover, there is often an almost point-to-point correspondence between dermal and epidermal melanophore patterns (Fig. 2–7). Although dermal melanophores are almost always

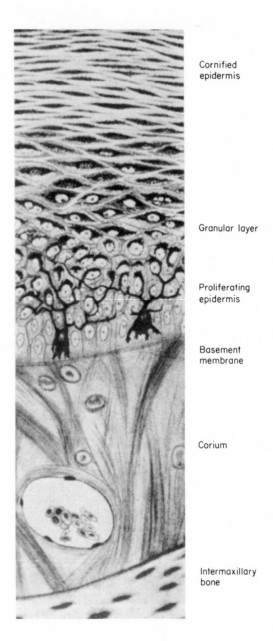

Cornified
epidermis

Granular layer

Proliferating
epidermis

Basement
membrane

Corium

Intermaxillary
bone

FIG. 2-5. Cross section of the bill of a sparrow illustrating cytocrine deposition of melanin within epidermal cells by melanophores. [From (7)]

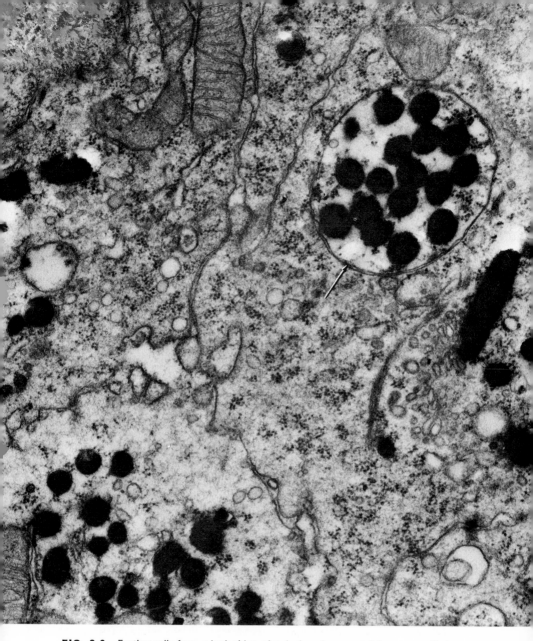

FIG. 2-6. Feather cells from a barb-ridge of a duck embryo containing an engulfed portion of an epidermal melanophore process. Melanophore cell membrane still intact (arrow). ×42,000. (Courtesy of Professor H. Koecke)

evident in poikilotherms, the occurrence of epidermal melanophores among these animals is highly variable. For instance, although there is little evidence that melanophores are prevalent in the epidermis of fishes, the presence of these cells in the reptilian epidermis seems the rule rather than the excep-

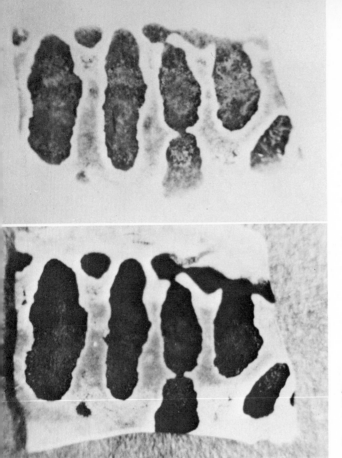

A

B

FIG. 2-7. Preparations of *Rana pipiens* skin illustrating patterns of pigmentation in split preparations of pure epidermis (**A**) and dermis (**B**). Note similarity in fine detail between dermal and epidermal pigment patterns.

tion. Among amphibians, epidermal melanophores are almost always found at some stage of embryonic, larval, or adult development. They may be absent from the adult, as in tree frogs (10), whereas the larvae usually contain these cells. The absence of epidermal melanophores from adult tree frogs is probably related to the fact that these species usually change color rapidly; the presence of epidermal melanophores with their associated cytocrine pigment would be an impediment to such changes.

From the standpoint of possible homologies between the melanophores of homeothermic and poikilothermic vertebrates, the nature of the epidermal melanophore deserves special recognition. It seems probable that if there is a very close relationship between the pigmentation of coldbloods and warmbloods, it resides at the level of the epidermal melanophore. First of all, as

FIG. 2-8. Premelanosome from the retinal pigment epithelium of the human fetus illustrating the highly organized structure on which melanin will be deposited. ×160,000. (Courtesy of Dr. A. Breathnach)

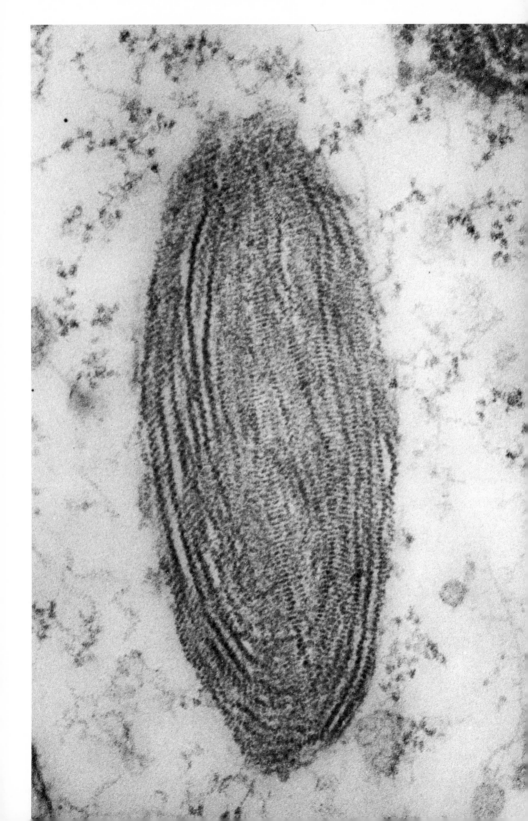

we have pointed out, epidermal melanophores of all vertebrate species are remarkably similar. At the same time, their appearance is clearly distinct from that of the dermal melanophores (Fig. 2–1A, B). Probably the most important activity of epidermal melanophores is their cytocrine function. Whether we consider the donation of melanosomes to the epidermis of frogs or man, to the feather of birds, or to the hair shaft of mouse or man, we are dealing with a function exclusive to the epidermal melanophore.

A fundamental feature of all melanophores is that they produce their own pigment—melanin. Melanin synthesis within melanophores involves, at least in vertebrates, the development of a subcellular organelle, the melanosome, upon which melanin has been deposited. Melanin is a complex polymer of tyrosine metabolites, the formation of which is catalyzed by tyrosinase. This enzyme is found on formative stages of the melanosome called premelanosome. The latter consists of a distinctive matrix of loosely coiled protein fibers on which the melanin polymers are deposited (11, 12). The premelanosome is formed from an intermediate vesicle of Golgi origin, approximately half a micron in diameter, that gradually assumes a more elliptical form (Fig. 2–8). After melanin formation is complete, the organelle is designated a melanosome. In other words, the term *melanosome* refers to what has been previously called a melanin granule. In order to avoid confusion, it should be pointed out that in the past the term melanosome had been used to designate one of the formative stages and the term premelanosome referred to an even earlier stage. According to current usage, all developing stages of the melanin granule are premelanosomes and the completely developed organelle is the "melanosome" (13). The melanosome is a more or less uniformly electron–dense organelle with no apparent ordered internal structure and with little or no tyrosinase activity. The term *melanin granule* is retained and may be used to describe all melanin-containing organelles that can be observed by light microscopy. For the most part, our knowledge of melanosome origin, formation, and activity is derived from studies on mammalian systems (11, 12). From limited observations of poikilothermic vertebrates, it appears that melanosomes in epidermal melanophores of these forms are much like those of mammals except for the fact that they are perhaps a little more spherical in shape (Fig. 2–9). The structure of melanosomes within dermal melanophores has not been systematically studied for any poikilotherms. However, scattered observations on various fishes, amphibians, and reptiles appear to be consistent with those made on mammalian epidermal melanophores. A notable exception occurs in the Mexican tree frog, *Agalychnis dacnicolor* (14). Melanosomes in dermal melanophores of this species are much larger than usual and are comprised of a dense internal core and fibrous cortex (Fig. 2–10). Moreover, these melanosomes are rust colored instead of the usual brown or black. Whether

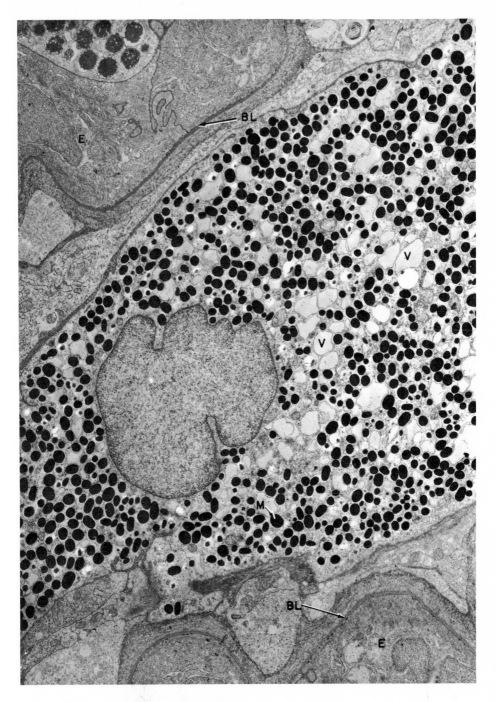

FIG. 2-9. Electronmicrograph of the skin of the Moor goldfish, *Carassius auratus*. A melanophore containing many melanosomes (M) and vesicles (V) is present in the dermis and is separated from the epidermis (E) on either side by the basal lamina (BL). ×7500. (Courtesy of Mrs. S. Holtan and Dr. J. D. Taylor)

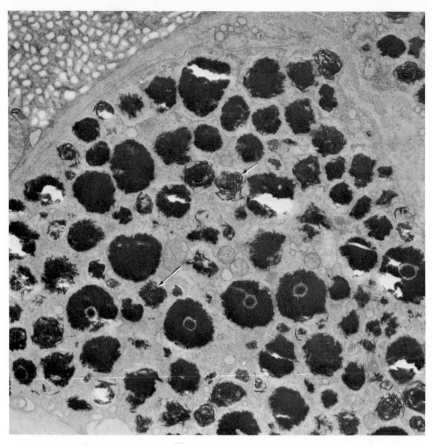

FIG. 2-10. Giant melanosomes within a dermal melanophore of the frog, *Agalychnis dacnicolor.* Note the fibrous nature (arrows) of the melanosomes. ×16,000. [From (14)]

melanosomes of *Agalychnis* represent a bizarre exception or are of a more common occurrence among amphibians can be answered only after more species are examined.

The majority of observations about melanophores presented so far pertain primarily to vertebrates, but, in general, some of these macroscopic features are also common to several invertebrates, especially those that change color readily, such as some crustaceans. Fundamental differences occur, however, with respect to the chemistry of the pigments they contain. Melanin is known to occur in melanophores of a variety of invertebrate types; yet its exact structural relationship to vertebrate melanins is unknown. Little is known about melanin-containing organelles of invertebrates. A common black or brown pigment of insects and crustaceans is ommochrome (15). This polymer of tryptophan metabolites is found in pigment granules that, in certain respects, resemble those within melanophores of vertebrates. Ommochrome is not a vertebrate pigment.

Iridophores

The reflection of light from pigmented surfaces of animals is often facilitated by the presence of cells called iridophores (16, 17). An important feature of these cells is that they contain organelles that are oriented in such a way as to reflect light efficiently. These cells often appear iridescent. Iridescence is not always accomplished by cellular structures; for example, the iridescent properties of feathers are not produced by intracellular pigments but, rather, are based on feather structure. In general, however, in most vertebrates and in many invertebrates, the reflection of light is accomplished by intracellular pigmentary structures. Whether such cells should be truly designated as pigment cells seems a legitimate question in view of the fact that the reflecting elements are not in themselves colored. For reasons that we hope will become obvious, it seems most reasonable to consider iridophores to be true chromatophores. This will be especially true for the vertebrate iridophores, which will be taken up first.

When viewed with reflected light, iridophores of fishes, amphibians, and reptiles appear silvery or golden. They are responsible for the metallic sheen that is often seen as flecks, spots, or stripes, or as broader areas such as ventral surfaces. Just as with melanophores, the pigmentary function of these cells is enhanced when their pigment is dispersed to the peripheral margins of the cell. When skins containing iridophores are viewed with transmitted light, their iridescence is not obvious; instead, they appear more or less opaque. This point is especially true when the pigment is concentrated toward the center of the cell. Under these conditions of illumination, the dispersed pigment often exhibits structural colors ranging from greens and blues in one portion of the cell to shades of pink or red in other areas (Frontispiece, J and K). Iridophores are generally restricted to the integument and they seem to be confined to the dermis. When their pigment is dispersed, iridophores are relatively flat and their processes extend out radially. Usually these processes are less dendritic than those of melanophores. It should be emphasized, however, that these cells are highly variable in appearance, depending on the species and on their specific anatomical location. Among tree frogs, for instance, the iridophores of the dorsal surface almost always appear punctate. Close examination reveals that these cells are filled with pigment and they appear round in shape (Fig. 2–11). Light reflected from the surface of these cells is subject to "Tyndall scattering," and, as a result, the cells have a bluish appearance.

It has been known for about a hundred years that the principal pigments of iridophores are purines, primarily guanine. It is now recognized that hypoxanthine, adenine, and uric acid may also be utilized as pigments. For many years it was known that these pigments are contained in rela-

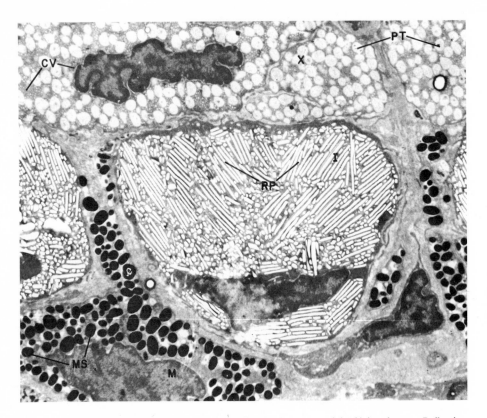

FIG. 2-11. Transverse section of skin from an adult *Hyla cinerea.* Reflecting platelets (RP) are oriented in stacks in the iridophore (I). Melanophores (M) contain melanosomes (MS). Xanthophores (X) contain pterinosomes (PT) and carotenoid vesicles (CV). ×5460. [From (18)]

tively large granules that impart a birefringence to the chromatophore. It has been possible to isolate these granules, and chemical analysis has revealed that they contain free purines. During the past few years studies with the electron microscope have revealed that, in iridophores of fishes, amphibians, and reptiles, these pigments are contained in organelles that appear as flat platelets. They been referred to as reflecting platelets (2, 16, 18) and are usually arranged in highly oriented stacks (Fig. 2–11).

FIG. 2-12. Lamellar-reflecting ribbons present within the iridophores of cephalopods. Ribbons cut frontally and transversely are shown. (Courtesy of Dr. R. A. Cloney)

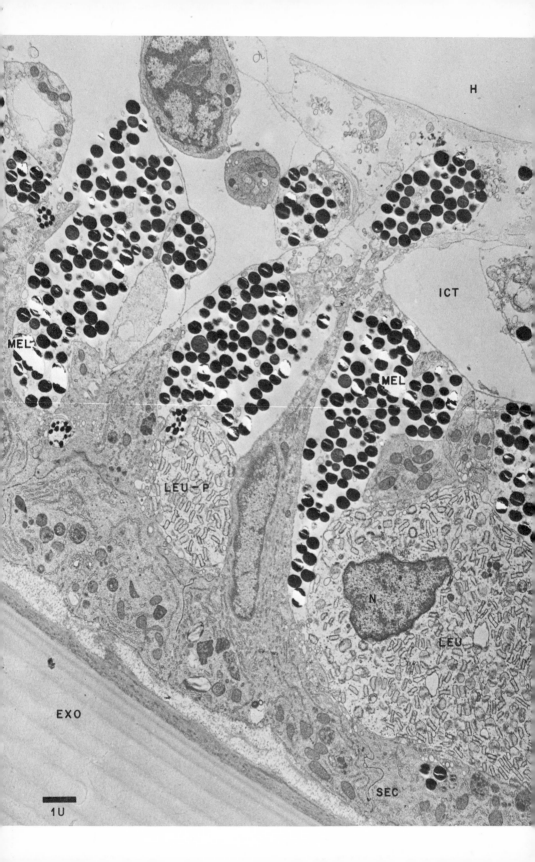

H

ICT

MEL

MEL

LEU-P

N

LEU

EXO

SEC

1U

Evidently this ordered arrangement of organelles is the basis for the reflective properties of iridophores. Relatively little is known about the nature of reflecting platelets other than the fact that they contain purines. Each platelet is bound by a membrane that presumably plays a dynamic role in the deposition of pigment, but little or no data are available about the origin of the platelets. Whether reflecting platelets disperse or concentrate during color change or whether an actual change in their size and shape occurs during this time are points that have not been established. It is likely, however, that both events may occur.

Iridophores are found most frequently in invertebrates and poikilothermic vertebrates; and although iridescent characteristics are sometimes associated with tissues of warm-blooded animals, normally they are not considered as being the result of iridophore function. For example, the eye shine that is seen at night in the eyes of many birds and mammals is not thought to be equivalent to the iridescence of the scales of fish. Recent evidence, however, suggests a change in point of view. This comes from observations of the golden-colored irises of some birds (19), which have revealed the presence of iridophores containing reflecting platelets much like those of frogs and fishes. Moreover, extracts of the irises from these birds are rich in guanine.

Iridophores are very common among invertebrates, but our knowledge of these cells is meager except for those species in which they have been studied recently at the ultrastructural level. The important feature of these invertebrate iridophores is that they contain some sort of highly organized system of granules or lamellae that serve to reflect light. In this respect, cephalopods have been studied most thoroughly, and it has been demonstrated that iridophores are present on the eye, the ink sac, and throughout the skin (17). These iridophores contain dense reflecting organelles that are derived from the Golgi and that project outward from the cell surface to form bodies composed of alternating reflecting surfaces and regular extracellular spaces (Fig. 2–12). The membranous organelles within adjacent iridophores are often oriented to one another to increase the size of the reflecting surface. The composition of the reflecting organelles is unknown. In a recent study of the skin of the crab, *Uca pugnax*, iridophores have been observed (20). It is fascinating that their granular organelles resemble the reflecting platelets of vertebrate iridophores (Fig. 2–13). Whether these structures also contain purines is not known.

FIG. 2-13. Fine structure of the fiddler crab epidermis showing leucophores (iridophores) and other integumental structures. EXO, exoskeleton; H, hemocoel; ICT, intraepidermal connective tissue; LEU, leucophore; LEU-P, leucophore process; MEL, melanophore; N, nucleus; SEC, secretory epidermal cell. ×9000. (Courtesy of M. R. Neff and Dr. J. P. Green)

Xanthophores and erythrophores

Pigment cells that are colored bright yellow, orange, or red are usually closely related chemically to one another. A simple terminology that has been adopted in order to designate these cells is based on their color; cells that appear yellow are referred to as xanthophores, and the term erythrophore is used if their color is red. Cells that are intermediate in color, such as orange, may be referred to as either xanthophores or erythrophores.

The role of these pigment cells in color change is quite variable among animals. There can be no doubt that the migration of pigments within xanthophores and erythrophores of crustaceans occurs following stimulation by hormones or other agents. The same seems to be true for many fishes. Among amphibians, these pigment cells play a more passive role; although important in imparting coloration to the skin, they are more or less static in morphology and are not normally involved in physiological color changes. There is only one instance in the literature describing pigment migration in xanthophores of an amphibian (21).

Xanthophores and erythrophores are very much involved in color patterns of vertebrates and are often found in discrete locations, such as in the red spots of the olive-backed red-spotted newt, *Notophthalmus virides-cens*. Although erythrophores and xanthophores are usually found in the dermis, the red spots of this animal are formed by localized concentrations of erythrophores in the epidermis. Newts generally have large accumulations of xanthophores and erythrophores on the ventral surface, leading to a "red-bellied" condition among some species of the genus *Triturus*. Often a manifestation of orange or red pigmentation is of nuptial significance, and such seems to be the case for the dewlaps of certain lizards, notably the anoles (22). In amphibians that change color rapidly, such as tree frogs, xanthophores are preferentially located on the dorsal surface where they are important in the establishment of green coloration. The significance of this point will be discussed later.

Except for the fact that the pigments of invertebrate xanthophores and erythrophores are predominantly carotenoid, relatively little is known about the nature of these cells. In contrast, our knowledge of the nature of vertebrate xanthophores and erythrophores has increased markedly in the past few years, particularly with respect to the kinds of pigments found in these cells.

Just as with the invertebrates, carotenoids are the major pigments of the xanthophores and erythrophores of fishes, amphibians, and reptiles. The old literature is so well documented in this respect that such pigment cells were often referred to as "lipophores," a fact that relates to the fat-soluble nature of the carotenoid pigments they contain. Hence it appeared almost heretical when Obika and Bagnara (23) reported in 1964 that

T A B L E   2 - 2

Analysis of pigmentation in adult skin of various amphibians

Species	Color of skin	Predominant chromatophore	Major colored pteridine	Solubility of pigment	Carotenoid test
Hemidactylium scutatum dorsal skin	Cocoa brown	Not studied	Sepiapterin	Not Studied	Negative
Plethodon cinereus dorsal skin	Metallic red	Erythrophore	Drosopterins	Soluble in H_2O, insoluble in fat solvents	Negative
Eurycea lucifuga dorsal skin	Dull red	Erythrophore	Drosopterins	"	Negative
Eurycea bislineata dorsal skin	Metallic yellow	Xanthophore	Sepiapterin	"	Negative
Pseudotriton ruber dorsal and flank skin	Brick red	Erythrophore	Drosopterins	"	Negative
Ambystoma maculatum dorsal yellow spot	Bright yellow	Xanthophore	Sepiapterin	"	Negative
Bufo punctatus (young adults) dorsal red spot	Bright red	Erythrophore	Drosopterins	"	Negative
Hyla arenicolor ventral and flank skin	Bright yellow	Xanthophore	Sepiapterin	Some in H_2O, some in fat solvents	Positive
Rana sylvatica ventral skin	Yellow-red	Xanthophore	Sepiapterin	"	Not tested

pteridines comprise much of the bright-colored pigments of fishes, amphibians, and reptiles and that some species utilize pteridines almost exclusively as their yellow or red pigments. (See Table 2–2.) About this time it became apparent that even for those species in which the adult pigment is primarily carotenoid, or for those others that utilize both pteridines and carotenoids, the first pigments to occur in the developing xanthophores and erythrophores are pteridines (24). In retrospect, something of the sort might have been suspected long ago, for it has been known for many years that animals cannot synthesize carotenoids and that the latter must be obtained from the diet. The same is true for riboflavin, which is apparently an important pigment among adult salamandrids (2).

Studies of the intracellular localization of these bright-colored pigments have been remarkably fruitful, for they have revealed that pteridines are contained in discrete organelles now known as pterinosomes (25). The

FIG. 2-14. Transverse section showing parts of a xanthophore and an iridophore in the skin of *Hyla cinerea*. Pterinosomes (PT) and carotenoid vesicles (CV) are found in the xanthophore, while the iridophore contains reflecting platelets (RP). ×4500. [From (10) and (18)]

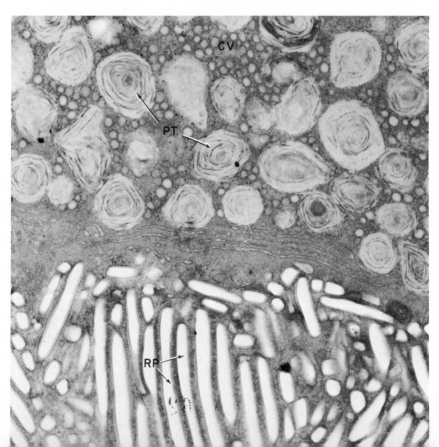

pterinosome is a spherical or ellipsoidal granule approximately half a micron in diameter and comprised of a series of concentric lamellae (Fig. 2–14). These organelles are distributed more or less uniformly in the cytoplasm, although in some erythrophores—for instance, those of the red swordtail fish (*Xiphophorus*) which contain both carotenoids and pteridines—the pterinosomes are essentially restricted to the periphery of the cell while the central area is occupied by carotenoid-containing vesicles. In xanthophores of other species—for instance, those of frogs—carotenoid-containing vesicles are distributed uniformly between the pterinosomes (Fig. 2–14). The origin of these vesicles is unknown, although it has been suggested that they derive from smooth endoplasmic reticulum. Pterinosomes have been observed in both xanthophores and erythrophores of fishes, amphibians, and reptiles, but thus far no real clues are available concerning the intracellular origins of these organelles.

It is well known that pteridines are important vertebrate pigments, and, in fact, these compounds were first isolated from the colored wings of certain butterfles (26). In modern times pteridines (especially the drosopterins and the sepiapterins) have been made famous by *Drosophila* geneticists who have studied pteridine patterns in the eyes of various geno-types. Ultrastructural studies of pigment granule development in the eyes of *Drosophila* have revealed the presence of pteridine-containing organelles that bear a remarkable resemblance to vertebrate pterinosomes (27). Not only is this pteridine-containing granule of *Drosophila* of a size comparable to those of vertebrates but internally its fibrous nature is not unlike that found in pterinosomes of some fishes. This is a rather interesting paral-lelism that warrants further study.

SPECIAL TYPES OF PIGMENT CELLS

Several other pigment cell types, usually seen in vertebrates, deserve mention. The *Langerhans* cell, which is found in the mammalian epidermis, contains no melanin; but on the basis of its general appearance as revealed by gold impregnation, it was formerly thought to be related to the epidermal melanophore (28). At the ultrastructural level, it is readily identifiable be-cause it contains peculiar rod-shaped organelles that have not been described in any other type of cell. At lower magnifications, the cytoplasm of these cells is clear; however, they should not be confused with "clear cells" (amelanotic melanophores). The latter term refers to epidermal melano-phores of certain genotypes wherein a failure of melanin synthesis leads to

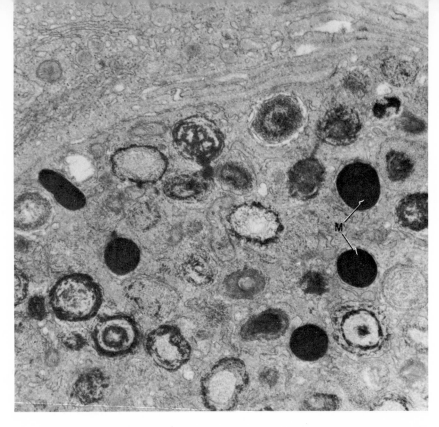

FIG. 2-15. An erythrophore of the dermis of an adult red-backed salamander, *Plethodon cinereus.* In addition to pterinosomes, melanosomelike organelles (M) are present. Note the electron dense regions (arrows) of some of the pterinosomes, suggesting intermediate forms between pterinosomes and melanosomes. ×20,000. [From (29)]

the formation of nonmelanized melanosomes as terminal products. Another melanin-containing cell of the integument is the *melanophage.* This is not truly a pigment cell but rather a macrophage that accumulates melanin granules incidentally as a function of its general phagocytic activities.

Ultrastructural studies of chromatophores have revealed that pigment cells of one type sometimes contain organelles characteristic of other chromatophore types. For example, melanophores containing pterinosomes have been found, and, conversely, xanthophores containing melanosomes have been discovered (Fig. 2–15) (29). Similarly, melanophores with reflecting platelets have been noted, as have iridophores containing melanosomes. Chromatophores containing both reflecting platelets and pterinosomes have also been found and pigment cells containing all three types of organelles may also exist. It is premature to attempt to classify these "hybrid" chromatophore types, for a great deal of thought should be given this problem before introducing such cumbersome terms as "melanoiridophore," and so on.

CEPHALOPOD CHROMATOPHORE ORGANS

Color changes in cephalopods (30) are mediated by remarkable structures that are, in reality, tiny organs consisting of five different cell types: the chromatophore proper (P), radial muscle fibers (MF), axons, glial cells, and sheath cells (SC) (Figs. 2–16 and 2–17). Pigment granules, either brown, red, or yellow, are confined to a compartment within the chromatophore. In the retracted state, the surface of the chromatophore is folded extensively (PI). These folds disappear as the chromatophore flattens during expansion. Expansion of the chromatophore results from a contraction of the radial muscle fibers, which leads to a stretching out and flattening of the sacculus to which the muscles are attached. During this process the diameter of the chromatophore increases some seven times. The radial muscles are associated with axons and glial cell processes, and adjacent muscle fibers are in contact with one another. The entire chromatophore and muscle fibers are surrounded by sheath cells. Color changes are effected by contraction of the radial muscle fibers of the various chromatophore organs. The chromatophores may be brown, red, or yellow, depending on the nature of the pigment granules contained in the pigment compartment. Although the colors of the different chromatophore organs are distinctive, their basic morphology is the same. The brown pigment-containing chromatophores are

FIG. 2-16. Diagram of a retracted squid chromatophore organ. [From (30)]

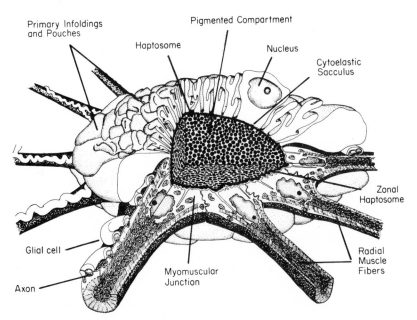

Primary Infoldings and Pouches

Pigmented Compartment

Haptosome

Nucleus

Cytoelastic Sacculus

Zonal Haptosome

Glial cell

Radial Muscle Fibers

Myomuscular Junction

Axon

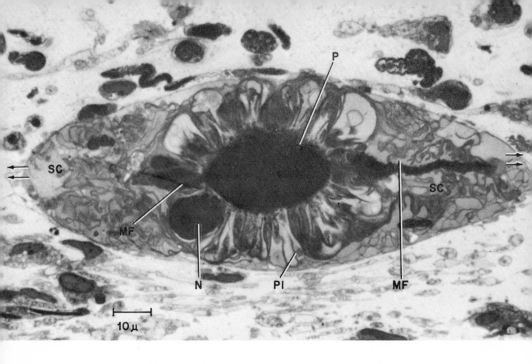

FIG. 2-17. Vertical section through a small retracted yellow-red chromatophore organ. The surface of the chromatophore is highly irregular and composed of primary infoldings (PI) and pouches. The nucleus (N) of the chromatophore lies outside the pigment-bearing compartment (P). Surrounding sheath cells (SC) and parts of two radial muscle fibers (MF) are visible. Double arrows indicate the plane of expansion of the chromatophore during contraction of the radial muscle fibers. [From (30)]

larger than those containing red or yellow pigment. At the ultrastructural level, the gross appearance of pigment granules of the various chromatophore types is similar.

PHYSIOLOGICAL COLOR CHANGES

Rapid color changes based on the intracellular mobilization of the pigment-containing organelles are referred to as physiological color changes. Usually this condition is a transitory response, and the organism can either revert back to its original coloration or it can assume an intermediate condition, depending on the stimulatory cues provided to the chromatophore. As will be pointed out later, a variety of physiological cues ranging from direct nervous stimulation to hormone action can produce physiological color change. In any event, the important consideration is that this is a rapid response ranging in duration from seconds to hours. Often these chromatophore responses follow a rather rigid rhythm, but more often they are the response to some immediate environmental stimulus, such as background change or alteration in conditions of illumination.

Perhaps the most common manifestation of physiological color change occurs with respect to background adaptation. In general, poikilotherms on a dark-colored background display melanophores with dispersed melanosomes, whereas the melanophores of animals on light-colored backgrounds possess melanosomes that are aggregated to a perinuclear position (Fig. 2–18A, B). Iridophores may also be important in physiological color changes, as demonstrated by the rapid dispersion and concentration of their pigments (Fig. 2–18C, D). It is thought that the eyes play a fundamental role in governing these background responses by reacting to the ratio of the amount of light falling directly on the eye to that which reaches the eye indirectly following reflection from the background. This ratio is referred to as the *albedo* and has been used extensively in studying the chromatophore physiology of fishes (31).

During the development of amphibian larvae there is a distinct point at which the ability to adapt to background is acquired. Prior to this point young larvae remain dark whether they are on light- or dark-colored backgrounds. Historically it was thought that these younger larvae which do not have the ability to background adapt were in a *primary phase,* whereas older larvae that can adapt were thought of as being in a *secondary stage.* In terms of chromatophore reactions, these two phases were considered sequential and clearly distinct from one another (1). The primary phase was regarded as representing a condition wherein larvae were darkly pigmented under conditions of illumination and were pale in darkness. It was assumed that this response does not involve visual function but rather depends on the reception of light by some other route. The secondary phase referred to what essentially constitutes adaptation to background or to the albedo, and it involves the reception of light by the ocular route. Our present state of knowledge confirms the existence of primary and secondary phases of color changes; however, the physiological basis for this distinction must be reevaluated. Based on all available evidence, it appears best to distinguish the primary phase as referring to those stages of larvae that have not yet acquired the ability to adapt to background, whereas the secondary phase should refer to older larvae or to adult stages that can adapt in response to background. Part of what was formerly considered the basis for the primary response—namely, the melanosome aggregation that occurs in darkness—is in reality not related to background response. It is considered chiefly attributable to effects of darkness on the pineal or, to a lesser degree, on the melanophores themselves. Moreover, this darkness–induced pallor or body-blanching reaction, as it is often referred to, persists in larvae that have long since acquired the capacity to background adapt and are clearly in the secondary phase. Onset of the secondary phase varies with species. For example, melanophores of *Xenopus* larvae respond to background changes

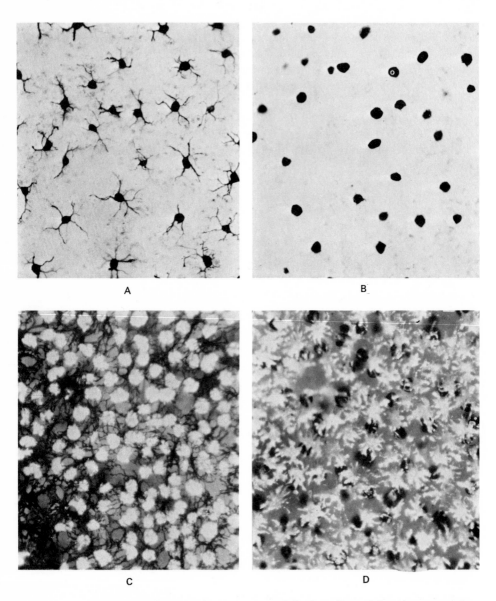

A

B

C

D

FIG. 2-18. Chromatophore responses of the frog, *Rana pipiens.* In response to a
black background (or to intermedin, *in vitro*), epidermal (**A**) and dermal (**C**) melanophores are dispersed. Melanosomes remain aggregated within these epidermal (**B**) and
dermal (**D**) melanophores in the absence of such stimulation. Iridophores (white-
colored cells, as seen here by reflected light), on the other hand, are punctate (reflecting
platelets are aggregated to the cell center) in response to intermedin stimulation (**C**)
but are dispersed in its absence (**D**). **A** and **B**, ×350; **C** and **D**, ×170.

from very young stages onward, but larvae of *Rana* and *Hyla* do not display the secondary response until they are several weeks old. In marked contrast, *Bufo* larvae may never acquire the secondary phase. The physiological basis for the acquisition of the secondary response will be taken up later.

MORPHOLOGICAL COLOR CHANGES

Color changes that are evoked slowly and that involve the accumulation or reduction in the amount of pigment contained in the integument fall into the category of morphological color change. Morphological color change is a slow process because it involves the synthesis or destruction of relatively large amounts of pigment as a result of either the persistence or lack of chromatophore stimulation. Usually background adaptation provides this cue, and therefore animals maintained on dark backgrounds develop more melanin while those on light backgrounds lose their melanin. The increase in pigment can result from the synthesis and subsequent cytocrine deposition of synthesized pigment as in the case of epidermal melanophores. During prolonged adaptation to dark backgrounds, frogs with epidermal melanophores deposit so much melanin in adjacent epidermal cells that, layer after layer, practically the whole dorsal epidermis becomes melanized. With respect to dermal melanophores, increased amounts of melanin formed during morphological color change are retained in the cell, and this increase is manifested in a denser appearance of the whole melanophore and its extended processes. A third expression of morphological color change results from a remarkable increase or decrease in the total number of pigment cells in the skin of the adapting animal (Fig. 2–19). Evidently such increases in numbers of melanophores are ascribable to a proliferation of existing melanophores or to the melanization of latent undifferentiated melanophores (melanoblasts). In any event, all these mechanisms lead to quantitative differences in the pigment content of the skin; this is the fundamental feature of morphological color change.

The examples chosen above pertain to the activity of melanophores in morphological color change. It should be emphasized, however, that other types of chromatophores are also active in this phenomenon; in fact, their activities may supplement those attributable to melanophores. This point has been shown rather nicely in studies of the role of albedo on pigmentation of fishes. For example, when animals are maintained on a white background, there is a diminution of melanin pigmentation and a concomitant augmentation of the iridophore pigment, guanine; conversely, on a black background the increase in melanin that ensues is accompanied by a proportional decrease in guanine content (31). In essence, the morphological effects elicited by the iridophore and the melanophore are supplemental to one another. Similar observations have been made on amphibians.

With respect to morphological color change following prolonged background adaptation, it should be pointed out that these changes are almost always preceded by physiological color change. In other words, the increase in the amount of pigment contained in a chromatophore seems to be related to the fact that the pigment-containing organelles of the cell are in a dispersed state, just as the decrease in pigmentary content is accompanied by an aggregation of these organelles to the cell center. This parallel relationship appears to hold not only for melanophores but for other chromatophores as well. It seems to be of such a general occurrence that some investigators have espoused the view that morphological color change is the necessary consequence of physiological color change. Some feel that increased synthesis is facilitated when pigment granules are dispersed because substrates are more readily accessible to enzyme sites on the granules, and they reason similarly that synthesis is inhibited when the pigment is aggregated because these sites are masked by the crowding of these organelles (32). Although these thoughts seem reasonable, and there is strong support for a causal relationship between physiological and morphological color changes, the universality of this relationship is in doubt because of several observations. A simple example of morphological color change in the absence of physiological change has been demonstrated in adult tree frogs, which contain dermal iridophores that are always in the aggregated state. Administration of intermedin to these frogs evokes a profound diminution of purines from these cells. The actual loss of this pigment is demonstrable both by direct measurement and by changes in the ultrastructural appearance of reflecting platelets in these cells (16). Another apparent factor involved in morphological color change relates to the increase in number of pigment cells that occurs after either prolonged hormone administration or maintenance on a dark background. It would appear that this manifestation of morphological color change is not necessarily the result of physiolgical color change but instead results from a stimulus of some sort on previously unpigmented cells (chromatoblasts). This situation probably occurred in the example shown in Fig. 2–19.

Although morphological color change is usually regarded as characteristic of poikilotherms, there are many examples of color changes in homeotherms that fall within the definition of this phenomenon. Most are concerned with seasonal alteration in the plumage of birds and in the pelage of mammals. These alterations are based on the relative amount of melanin pigment that is eventually deposited in the growing feathers or hairs. Usually these cyclic changes reflect quantitative differences in the amount of melanin pigments synthesized by the epidermal melanophore.

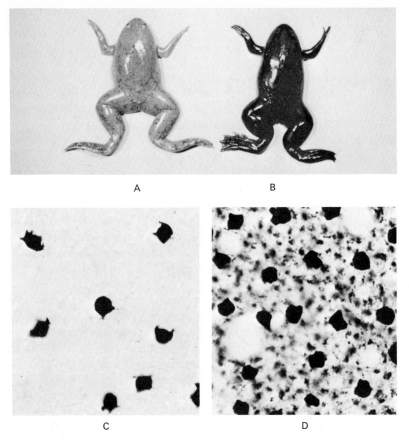

FIG. 2-19. Two adult *Xenopus* that have been kept on a white (**A**) and a black (**B**) background, respectively, for six weeks. Skin preparations beneath indicate that the white-background-adapted frog has few epidermal melanophores (**C**), whereas the black-background-adapted frog has many melanophores (**D**). The melanophores in both preparations are punctate because the skins were maintained in Ringer solution prior to fixation. Note cytocrine melanin within the epidermal cells from the skin of the black-background-adapted frog. **C** and **D**, ×430. [From (6)]

However, qualitative differences resulting from the synthesis of different kinds of melanins, by the same chromatophore, should also be considered as examples of morphological color change.

DISPERSION AND AGGREGATION
VERSUS EXPANSION AND CONTRACTION

There has never been much doubt that the participation of individual chromatophores in physiological color change is concerned with the spreading of pigment over a larger or smaller surface. The question of how this process is accomplished, on the other hand, has been controversial for many years (1). Originally the terms expansion and contraction were used to designate, respectively, the states where more surface or less surface were covered by the pigment cell. For a time this terminology implied that pigment cells actually change their shape so that their diameters are considerably greater in the expanded state than in the contracted or punctate state. This concept was challenged early by the alternative concept that chromatophore boundaries are fixed and that migration of pigment within the chromatophore actually is responsible for the conformational changes witnessed during physiological color change. The principal evidence cited to support this concept relates to the fact that after repeated responses individual pigment cells always display the same profile. Proponents of the "expansion-contraction" concept countered with the explanation that "expansion and contraction" of the chromatophore indeed occurred and that a final static profile was maintained because pigment cells were imbedded in a matrix containing permanent channels into which the expanding chromatophore dendrites were forced to return. Observations made during the past thirty years present overwhelming evidence that, with respect to most chromatophore types, physiological changes involve a migration of pigment granules within a cell that has fixed boundaries. The sole exception occurs in melanophores isolated in tissue culture and even here the degree of expansion or contraction of these cells is minimal (33). Notwithstanding the general argument that pigment granules migrate within the chromatophore, the terms expansion and contraction are still in use. This situation is probably due to the fact that it is more unwieldly to refer to migration of pigment within the cell than it is to refer simply to the expansion or contraction of the cell, but in part it may relate to the relative lack of knowledge about some pigment cell types. For example, it is not really known what happens to the reflecting platelets of iridophores during physiological color change. It seems most likely that reflecting platelets are somehow displaced to and from the periphery, but the possibility of the occurrence of conformational changes of the chromatophore itself has not been eliminated. Similar questions apply to xanthophores and erythrophores.

STRUCTURAL COLORATION

The contribution of physical structures to the pigmentation of animals is extremely important, as can be seen by observing the bright colors of such unrelated organisms as beetles and birds. This area of structural coloration consists of three principal sources: interference, diffraction, and Tyndall scattering. Because our treatment of pigmentation is oriented primarily toward physiological control of color changes, our major emphasis is on Tyndall scattering, for it is this phenomenon that is primarily responsible for the blue or green coloration seen in amphibians and lizards, which we know change color rapidly. Tyndall blues may result from the scatter of light from iridophores. When light strikes the reflecting platelets within these cells, there is a differential scattering of light such that the short light rays are diffracted while the slow long rays pass through. Consequently, iridophores may appear bluish in reflected light except when they are overlain by yellow pigment cells, which, in a sense, act as yellow filters to provide green coloration. When viewed with transmitted light, iridophores may appear as mixtures of green, yellow, orange, or red, depending on the penetration of longer wavelengths of light. The density or orientation of reflecting platelets of some iridophores is such that the Tyndall blues of partially reflected or scattered light are replaced with the whites of totally reflected light. When such iridophores are overlain by xanthophores, the greenish appearance is replaced by bronzes or golds.

ASSESSMENT OF
CHROMATOPHORE RESPONSES

In dealing with the action of various physiological agents on chromatophores, it is important to have some criterion by which the degree of response can be evaluated. Consequently, a variety of methods have been developed over the years, some of which are macroscopic and involve attempts to assess the general shade of animals. Because of the obvious frailties of such subjective schemes, other systems were developed and, finally, the subjective but quantitative method of Hogben and Slome (34) appeared and achieved wide usage (Fig. 2–20). In essence, this method

FIG. 2-20. Stages comprising the melanophore index of Hogben and Slome (34).

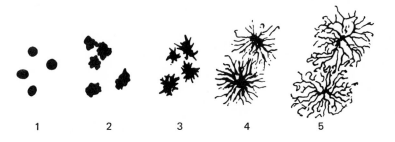

1　　　　2　　　　3　　　　4　　　　5

employs five distinct stages representing different degrees of pigment dispersal within the chromatophore: the most aggregated state is designated as stage 1 and the most dispersed as stage 5. Stages 2, 3, and 4 represent intermediate degrees of pigment dispersal. Verbal designations of these stages are: 1, punctate; 2, punctate-stellate; 3, stellate; 4, stellate-reticulate; 5, reticulate. By using this system, an index of the relative dispersal of chromatophore pigments can be obtained for quantitative use or for graphic representation. This method is commonly used in assessing melanophore responses; hence the term *melanophore index* or MI appears in the literature frequently. There is no reason why this method should not be used for other types of chromatophores that undergo the same degrees of pigment dispersal in their physiological responses. The Hogben and Slome method has received some criticism because of the fact that it is basically subjective. Although this point is true, its use has been of singular importance in leading to a number of important physiological discoveries. Moreover, those laboratories that continue to use this method boast of reliability that compares favorably with some of the less subjective procedures. Perhaps the greatest disadvantage of this chromatophore index method is that it is time consuming.

Photoelectric methods of assessing chromatophore responses are currently in vogue; there are several modifications but essentially they involve measurement of the fraction of incident light that is reflected from a unit area of pigmented skin surface. These methods have the advantage of being objective and rapid. Moreover, since the most widely used reflectance assay method is an *in vitro* system (35), the problem of indirect pigmentary responses is obviated, and, at the same time, it becomes possible to study the action of substances that might be toxic when administered at the organismic level. The *in vitro* technique offers the additional advantage of testing the action of several substances administered either simultaneously or in sequence. The fundamental disadvantage of this reflectance assay system is that it is not possible to ascertain effects at the level of individual chromatophores. This fact is important, for the reflectance changes from the skin of a frog following the administration of a specific agent are at least the resultant of effects on both melanophores and iridophores (36). As long as both pigment cells are affected in a parallel way, there is no real problem. However, differential action in which one chromatophore type is stimulated and the other is not could lead to erroneous interpretations. This situation can be controlled, however, by subsequent cytological observations of the pigment cells in either living or fixed preparations of skins. Depending on the specific pigmentary problem being studied, either the skin reflectance method or the melanophore index method is a reliable means of assessing chromatophore responses.

CELLULAR ASSOCIATIONS

During the past few years studies of functional associations between pigment cells, or between pigment cells and neighboring epidermal cells, have made valuable contributions to our knowledge of pigment cell biology. Two important concepts have emerged that display, rather nicely, unified functions of these cellular associations: the *dermal chromatophore unit* (10) and the *epidermal melanin unit* (5). The former is a primary vehicle for physiological color change among poikilotherms and the latter is concerned primarily with morphological color change of both poikilotherms and homeotherms. In addition to these two concepts, which are important to pigmentation of poikilotherms, other cellular associations significant in the expression and the enhancement of patterns of other forms will be considered.

The dermal chromatophore unit

That dermal chromatophores are the important pigment cells of vertebrates that color adapt rapidly has been known for many years. Moreover, it is well established that in the dermis of these animals, including most notably tree frogs and lizards, there is a specified localization of the three chromatophore types: xanthophores are uppermost and are found just beneath the basal lamina and associated collagen; iridophores form a layer or layers below the xanthophores; and the most basal of the pigment cells are dermal melanophores (Fig. 2–21). The latter may be found either in very close association with the overlying iridophores or they may be even more deeply located in the dermis. At any rate, it is fundamental that dendritic processes extend upward from these melanophores to terminate above the iridophores. In amphibians, the iridophore layer is only one cell thick and here melanophore processes terminate on the upper surface of the iridophore, between it and the overlying xanthophores (Figs. 2–21 and 2–22). Together these three distinct chromatophore types comprise a functional unit (*dermal chromatophore unit*) that is responsible for bringing about rapid color changes. It should be emphasized that although the unit contains three varieties of pigment cells, this does not mean that these different chromatophores are present necessarily in a one-to-one relationship. The concept of this unit is a functional one that includes a layering or a grouping of pigment cells in such a way that while each variety of chromatophore is represented, one type may be more prevalent than another. The relative numbers of each chromatophore that is present in dermal chromatophore units vary with the species being considered: That of *Rana pipiens* contains relatively few melanophores in comparision to the numbers

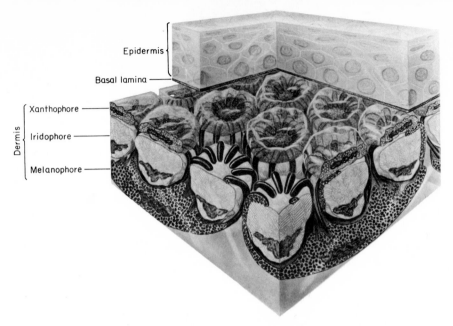

FIG. 2-21. Schematic interpretation of the dermal chromatophore unit from several anurans. Adaptation to a dark background is represented. [From (10)]

FIG. 2-22. Transverse sections through the skin of adult *Agalychnis dacnicolor* **A.** Light background. Melanosomes are restricted to the perinuclear area of the melanophore. Iridophores and xanthophores are visible above each melanophore. **B.** Melanosomes are dispersed during adaptation to a dark background and move from the perinuclear area to occupy processes between the xanthophore and the iridophore. Compare with **A.** ×1000. [From (18)]

A   B

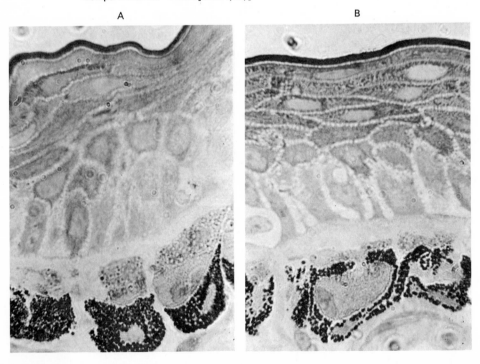

of iridophores and xanthophores that are present (Fig. 2–18C, D). In contrast, the unit found in *Hyla arenicolor* contains many fewer xanthophores than other chromatophore types. The dermal chromatophore unit of the dorsal surface of *Agalychnis dacnicolor* is unusual in that it is composed of one of each of the three chromatophore types (Fig. 2–22). Although the first functional description of the dermal chromatophore unit was made from studies of amphibians, there is no reason for not extending the use of this concept to lizards (37), specifically in view of its basic similarity to a scheme proposed by von Geldern (38) to explain color change in the Carolina fence lizard, *Anolis carolinensis* (Fig. 2–23).

The following material is a brief description of the functioning amphibian dermal chromatophore unit. During adaptation to light-colored backgrounds when the circulating levels of pituitary chromatophorotropic hormone are low, melanosomes are aggregated to a perinuclear position and the melanophores contribute little to the color of the animal (Fig. 2–22A). At the same time iridophores are manifested fully; first, because they are not obscured by any overlying melanin and, secondly, because of the fact

FIG. 2-23. Schematic interpretation of the dermal chromatophore unit of *Anolis carolinensis*. Note that melanophore processes terminate above the xanthophores. Melanosomes are shown in various states of intracellular distribution. [From (37)]

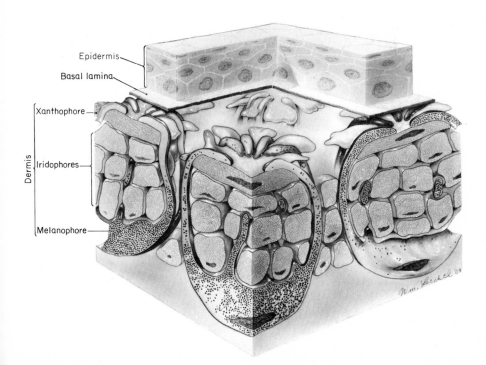

that in those iridophores that show physiological color change, reflecting platelets are fully dispersed. The net effect of these complementary melanophore and iridophore responses is that the animal becomes much lighter in color. The third component of the dermal chromatophore unit—namely, the xanthophore—makes a passive contribution by serving as a yellow filter or screen. Xanthophores in this unit apparently do not show physiological color change and thus make no dynamic contribution in the function of the unit. The role of the dermal melanophores is seen during adaptation to dark-colored backgrounds when circulating levels of chromatophorotropic hormone apparently rise to a level sufficient to stimulate the migration of melanosomes into the distal-most melanophore processes. At the same time the aggregation of iridophore pigments takes place (in iridophores that so respond), thereby leading to a reduction in the reflecting surface area of the iridophore. The effectiveness of this reflecting surface is further diminished by the profound masking produced as melanosomes fill the fingerlike melanophore processes that cover the iridophore (Fig. 2–24). Intermediate degrees of darkening are attributable to a submaximal response of both iridophores and melanophores.

FIG. 2-24. Transverse section through the skin of an adult *Hyla cinerea* treated with MSH. A fingerlike melanophore process (M) (cut transversely) containing melanosomes is seen between the upper surface of an iridophore (I) and the lower surface of a xanthophore (X). ×30,000. (Courtesy of Dr. J. D. Taylor)

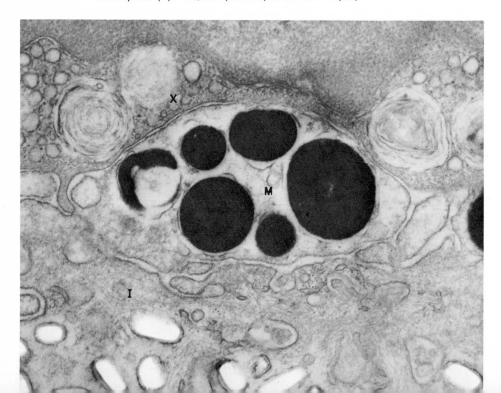

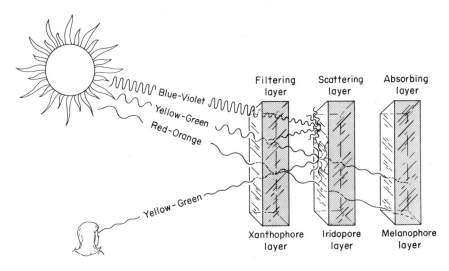

FIG. 2-25. Diagramatic interpretation of the basis for green coloration in vertebrates (For explanation, see text.)

The description just presented explains the functional integration of the three chromatophore types in darkening or lightening of the animal, but it does not explain how the unit contributes to color change, particularly as frogs or lizards change from brown to green or white. In order to explain this process adequately, it is necessary to integrate structural coloration into the concept of the dermal chromatophore unit. The iridophore is involved specifically at this point, for it is this cell that either totally reflects incident light to produce whiteness or partially scatters light to produce Tyndall blues. In the first case, it is assumed that the reflecting platelets of the iridophores are either densely packed together or are oriented so that all wavelengths of incident light are reflected back in the same proportion. In this way white light is reflected from the iridophore surface to impart whiteness to the animal (15). If xanthophores are present above the iridophore layer, violets and blues are removed from the spectrum of this reflected light by the carotenoid or pteridine pigments and the animal appears yellowish. When the arrangement of reflecting platelets in the iridophore is such that incident light is only partly reflected, a Tyndall effect is produced, leading to a blue appearance. In this case, the longer wavelengths of light— red and orange—are transmitted through the iridophore layer and are ultimately absorbed by the underlying melanophores while the shorter wavelengths, dominated by the blues and greens, are reflected. Again, the covering xanthophore layer absorbs the blue fraction of these wavelengths so that, finally, greens are primarily reflected from the surface of the animal (Fig. 2–25). It is interesting that mutants in which yellow pigment is either reduced or missing from the xanthophore layer are blue in color.

An interesting parallel of the dermal chromatophore unit leads to the establishment of green coloration in the feathers of some birds. Although these feathers do not possess the same chromatophore types that are involved in the amphibian or reptilian dermal chromatophore unit, the same three functional elements—a filtering layer, a scattering layer, and an absorbing layer—are present in another form (39). In the feather shaft, for instance, there is a deposit of melanin granules that is proximal to a keratinized cell layer. The keratinized layer provides the blue coloration, probably a Tyndall effect, which is enhanced by the absorption of longer wavelengths by the deeper melanin granules. Distal to the keratinized layer in the outermost portion of the feather shaft, carotenoid pigments are found either as droplets or in a dispersed state. The net result is that the bluish coloration of deeper layers is translated to green because the shortest wavelengths are filtered out as light passes outward through the yellow carotenoid layer. Just as in the case of amphibians and reptiles, blue mutants occur when the yellow pigment is deficient or absent.

The epidermal melanin unit

Epidermal melanophores are the most widely distributed of all vertebrate pigment cells, and whether in frogs or humans, their general morphology and function appear similar. Perhaps the most fascinating characteristic of this pigment cell is that it carries out its pigmentary function in association with other cells. This step is accomplished by the transfer of melanin pigment from the epidermal melanophore, where it is formed, to the surrounding or adjacent pool of Malpighian cells (Fig. 2–26). These epidermal cells, together with an epidermal melanophore, make up the *epidermal melanin unit*. The transfer of melanin serves to emphasize that this is a unit of morphological color change that is expressed not only in the epidermis but in associated structures as well. Many of the color changes that occur in the bills and feathers of birds and in the pelage of mammals result from the deposition of melanin into the epidermal cells that give rise to these specific cornified structures. These examples are particularly significant because they amplify the precise nature of the association between the epidermal melanophore and the epidermal cells served by this chromatophore. The existence of such an exact association must be inferred from the great uniformity in pigment deposition that occurs during pigmentary events, such as the formation of either barred feather patterns or agouti coats.

The fact that the epidermal melanin unit comprises two discrete elements—the epidermal melanophores that produce this pigment and the epidermal cells that act as pigment receptors—raises interesting questions relative to the "communication" that obviously exists between these elements (40). First, how is the initial contact between these cells established? Does

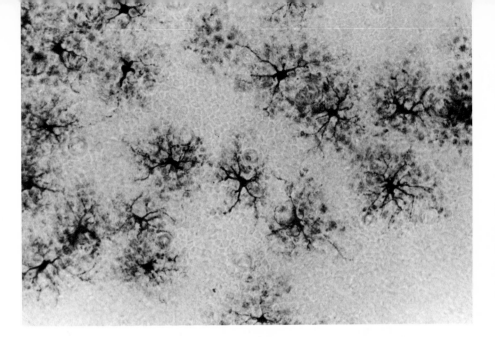

FIG. 2-26. Discrete *epidermal melanin* units having little or no functional overlap as seen in the epidermis from the adult frog, *Rana pipiens.* The pattern of epidermal. melanization (cytocrine activity) is correlated with the degree to which melanosomes are dispersed within the dendritic processes of the epidermal melanophores.

the newly arrived epidermal melanoblast select the epidermal cells it is to make contact with, or is this role assigned the epidermis, which somehow "attracts" dendrites from appropriate melanoblasts? The question of physiological "communication" between elements of the epidermal melanin unit is also important and thought provoking. For example, although we tacitly assume that pigment-cell-stimulating hormones operate at the level of the melanophore, it is also possible that these hormones may affect the epidermal cells, which, in turn, lead the epidermal melanophores to synthesize and deposit melanin. In other words, we cannot, at the moment, completely exclude the possibility that there are at least two inputs into the system, one at the level of the epidermal melanophore and the other at the level of the epidermis.

PATTERNS

Both the dermal chromatophore unit and the epidermal melanin unit represent dynamic pigment cell associations that have an immediate functional significance in the physiology of the organism. In both of these cellular associations, intimate cellular contact is the rule. In contradistinction, an analysis of the basis of some of the various color patterns that exist among amphibians reveals that while definitive cellular associations are involved,

43

the existence of these patterns is not necessarily of immediate functional importance to the animal. Moreover, critical histological examination discloses that cellular associations in some patterns are often at considerable distance from one another. A particularly clear example of the latter point is demonstrated by a comparison of the relative distribution of epidermal melanin units and dermal chromatophore units in a frog such as *R. pipiens* that has a definitive spot pattern. In the large black spots of the animal, epidermal melanin units are both numerous and well developed, whereas in the dermis the dermal chromatophore unit is not fully formed. The edges of these spots are marked precisely by an absence of epidermal melanin units, together with the appearance of well-developed dermal chromatophore units (Fig. 2–27). Throughout the integumental areas between the spots, a point-to-point correspondence exists between the absence of epidermal melanin units and the presence of dermal chromatophore units. Although it is indeed remarkable that dermal and epidermal chromatophore patterns can be so mirrored, it seems that this type of association is fairly common. Another example of this dermal-epidermal relationship is provided in the red spots of adult *Notophthalmus viridescens,* the olive-backed or red-spotted newt. The red spots are composed of epidermal erythrophores (containing carotenoid pigments) that are distributed in exact correlation with a discrete dermal spot composed of iridophores. When the whole skin is

FIG. 2-27. Transverse section of the skin of an adult *Rana pipiens* at the edge of a spot. Spot area on the left is marked by the presence of large amounts of epidermal cytocrine melanin. No dermal chromatophore units are found beneath this area. To the right, the clear epidermis is underlain by well-developed dermal chromatophore units.

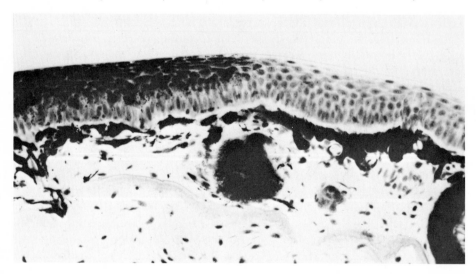

leached in alcohol, what was a red spot becomes a bluish spot because the iridophore component in the dermis is retained while the contents of the erythrophore portion of the dermis is dissolved.

The association of iridophores with overlying xanthophores or erythrophores is a frequent occurrence among animals that have a definite spot pattern. The advantage of such an association is apparently related to the reflective properties of the iridophores in that the color of the overlying chromatophore can be enhanced by light reflected through it from the iridophore surface. Among poikilotherms, individual iridophores and xanthophores are seen together so frequently that the term *xantholeucophore* was introduced to designate this association (1). Often, however, the intimate association of these two cell types is merely fortuitous, and, as a result, it seems difficult to understand just when this term is to be used. Accordingly, we have purposely excluded this term from our section on terminology. Although the use of xantholeucophore is of questionable value in designating a specific morphological structure, it is indicative of a principle that is far more widespread in its occurrence than is illustrated by the few vertebrate examples mentioned above. Among invertebrates, cephalopods stand out in this respect, for in this group it is frequently seen that chromatophores found at superficial levels of the dermis are underlain by more deeply located iridophores. This is indeed an interesting parallelism, for the cephalopod iridophore is, of course, in no way related to the vertebrate iridophore except that its final function to reflect light is the same.

Cellular associations in the establishment of chromatophore patterns are obviously complex. The few examples presented above serve to show the kinds of associations that exist and to imply that complex processes are involved in the establishment of these patterns.

CHAPTER

3

biochemical aspects
of chromatophore pigments

The principal pigments of animals include melanins, ommochromes, purines, pteridines, carotenoids, and flavins. Such an array of different compounds as represented by these pigments is difficult to conceive, for, among vertebrates, they are all derived in or are accumulated in cells originating, in common, from the neural crest. Even more fascinating is the fact that invertebrates and vertebrates alike, in a nonhomologous way, have developed the use of melanin, purines, pteridines, and carotenoids as fundamental pigments. Consequently, this discussion will center around these more commonly occurring classes of compounds, although it should be pointed out that a variety of other pigmentary substances occur frequently and at random among animals.

MELANINS

Melanin (*melanos*: black) is a generic term used to designate tyrosine-derived pigments of high molecular weight and great stability. Although most are darkly colored, lighter-colored melanins are often encountered. Generally black or brown pigments are referred to as eumelanins, whereas yellow or orange melanins are termed phaeomelanins.

In keeping with their wide distribution, melanins have commanded more attention than other classes of pigments, and, as is so often the case, much of our early knowledge of this subject was gained from early studies made on plants. Just before the turn of the century, it was learned that an enzyme present in a mushroom was capable of converting a colorless substrate found in this organism to a black pigment. The black product, of course, was melanin. Subsequently, when it was learned that the substrate was tyrosine, the enzyme was named tyrosinase. The presence of this copper-containing enzyme in animal tissues was revealed soon afterward; however, much of the early work on melanin biosynthesis utilized plant materials. Consequent to extensive studies made in the 1920s by Raper and his collaborators, with the use of plant tyrosinases, a metabolic scheme for the synthesis of melanin was proposed (41). This general concept of melanogenesis has survived over the years, although there are currently indications that final steps in the scheme need clarification (42). Possibly this scheme is slightly variable, depending on the animal being studied. For our purposes, the general pathway presented in Fig. 3–1 is acceptable; however, it should be kept in mind that the question of the nature of the polymerization involving indole-5, 6-quinone is still unresolved. It has been considered that eumelanin is a homopolymer composed of units of indole-5, 6-quinone (42), but other work (43) based on extensive chemical degradation procedures supports the view that eumelanin is a heteropolymer made up of indole units at different levels of oxidation and other tyrosine metabolites. The lack of indisputable evidence concerning the chemical configuration of the eumelanin polymer is due to the fact that it is so stable that only radical chemical reactions degrade it. As a result, chemical studies are difficult to perform and results are difficult to assess. The great stability of this pigment raises important biological questions concerning the disappearance of melanin in the living system. Surely molting or the gradual loss of skin could account in large measure for the loss of melanin from the integument; however, the fact that cytocrine melanin may exist in packets resembling some lysosomes (Fig. 2–6) raises the possibility of enzymatic degradation. If such an enzymatic breakdown occurs, it is possible that it relates to a destruction of the protein component of melanoprotein rather than to the eumelanin polymer itself.

Phaeomelanins have been the recent subject of particularly important studies (44) showing that these compounds differ from eumelanins not only in their yellow-orange color and their solubility in dilute alkali solutions but in their composition as well. Elementary analyses of phaeomelanins derived both from the red feathers of chickens and the red hair of man have revealed the presence of sulfur. Subsequently, it was shown that the oxidation of equimolar mixtures of dopa and the sulfur-containing amino acid cysteine leads to the formation of a yellow-orange compound much like phaeomelanin. Treatment of this compound with mineral acids produced a pigment identical to a reaction product obtained from naturally derived phaeomelanin. These findings have led to the suggestion that naturally occurring phaeomelanins are formed from alterations in the eumelanin

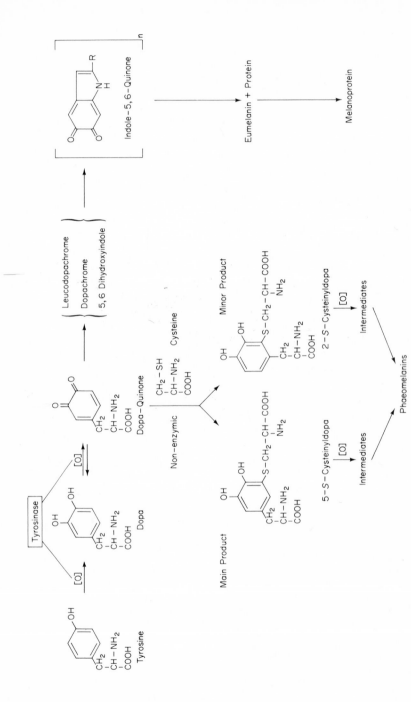

FIG. 3-1. Biosynthetic pathways for phaeomelanins and eumelanins from tyrosine within melanophores.

pathway, involving an interaction of cysteine with quinones derived from the oxidation of dopa (3,4 dihydroxyphenylalanine) to form 5S–cysteinyldopa and 2S–cysteinyldopa in a molar ratio of 95:5. These products are then further oxidized to form phaeomelanins (Fig. 3–1). This suggested pathway of phaeomelanin biosynthesis has recently been supported by the structure elucidation of a group of phaeomelanic pigments (45) previously called trichosiderins (Fig. 3–2). These pigments are widely distributed among a variety of mammalian and avian species possessing yellow, orange, or red hair or feather coloration. It should be emphasized that light-colored pelage or plumage does not necessarily signify the presence of phaeomelanins.

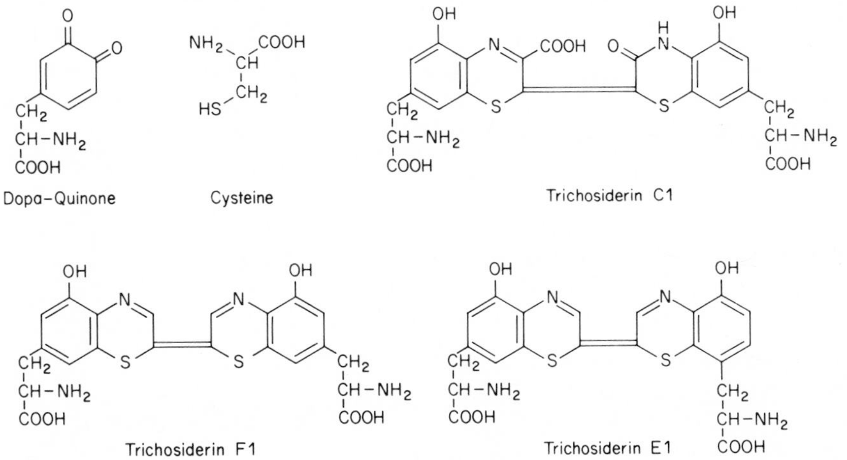

FIG. 3-2. Three naturally occurring trichosiderins derived from the reaction of cysteine and dopa-quinone.

The identity of these pigments must be confirmed by chemical analysis, for even eumelanins may impart reddish or brownish coloration, depending on how they are ultimately distributed in the skin, hair, or feathers.

It is an interesting fact that phaeomelanin may occur in animals that also produce eumelanin. The agouti coat color of mice is a particularly good example, for it appears that the dark portions of the hairs represent periods of eumelanin synthesis, whereas the lighter-colored regions are produced during periods of phaeomelanin synthesis. Apparently an individual melanophore possesses the capability of synthesizing both eumelanin and phaeomelanin and does so under the appropriate stimuli. Although the possible interconversion of eumelanin and phaeomelanin synthetic pathways has not been studied in man, it may exist and could thus explain the gradual reddening of hair and skin of certain Negroids during periods of dietary insufficiency. The significance of phaeomelanin as a vertebrate pigment is only now being revealed, and combined chemical and biological studies of this pigment class promise to be interesting.

OMMOCHROMES

Ommochromes constitute a group of brown, yellow, or red pigments that are often confused with melanins. In cells, they occur in granules that resemble melanosomes, both in size and general appearance. They are relatively stable, insoluble in water, and are often found in animals that also synthesize melanins. Unlike melanins, however, ommochromes are formed from the oxidation of tryptophan. The name ommochrome is derived from the association of these pigments with the ommatidia of arthropod eyes. Although ommochromes are known to occur in a variety of invertebrates, they appear to be restricted to protostomes, having been found in arthropods, echiuroids, and mollusks. Most of our knowledge of these pigments is derived from work on insects, notably dipterans and lepidopterins. Ommochromes are not found in vertebrates.

Thanks to chemical experimentation carried out largely by Butenandt (46), we know that ommochromes fall into two classes, the ommatins and the ommins. Ommatins are low-molecular weight, readily dialysable, and alkali-labile compounds, while ommins are high-molecular weight, nondialysable, and alkali stable. That both of these ommochrome classes are derived from tryptophan was shown from experiments in which the C^{14}-labeled amino acid was found subsequently in extracted ommatins or ommins. The chemistry of ommins is relatively complex: therefore more work has been done on the ommatins, leading finally to the isolation of both xanthommatin and rhodommatin and to the structural characterization of the former. The integration of genetic studies on eye pigmentation of *Drosophila* with later chemical investigations has led to our understanding of the initial steps of ommatin biosynthesis (Fig. 3–3). In the first step, tryptophan is coverted to kynurenine by an enzyme under the control of gene v^+. The second step, controlled by a product of gene cn^+, involves the oxidation of kyurenine to 3-hydroxykynurenine. Steps in the subsequent oxidation and condensation of two molecules of 3-hydroxykynurenine lead to the formation of various ommatin pigments, such as xanthommatin. A further genetic involvement in ommochrome pigmentation concerns the fact that ommochromes, like melanins, are also attached to proteins. Activation of genes controlling enzymes that function in the attachment of the pigment to the protein carrier is an important factor in the ultimate expression of pigmentation. In the course of genetic studies on eye pigmentation of *Drosophila,* a possible relationship between red and brown eye pigments has been suggested, even though the brown pigments are ommatins and the red pigments are pteridines. Since these compounds are chemically unrelated, it is difficult to envision any direct transformation; however, it has been suggested that they might compete with one another for common precursors (47). Another possibility is that one or more of the pteridine components serve as a cofactor in the pathway of ommochrome synthesis. This concept is reminiscent of melanin synthesis wherein a pteridine cofactor is involved in the hydroxylation of phenylalanine to tyrosine (48). Altogether, our understanding of ommochrome chemistry is in its early stages, and it is obvious that this is a fertile area for new discoveries.

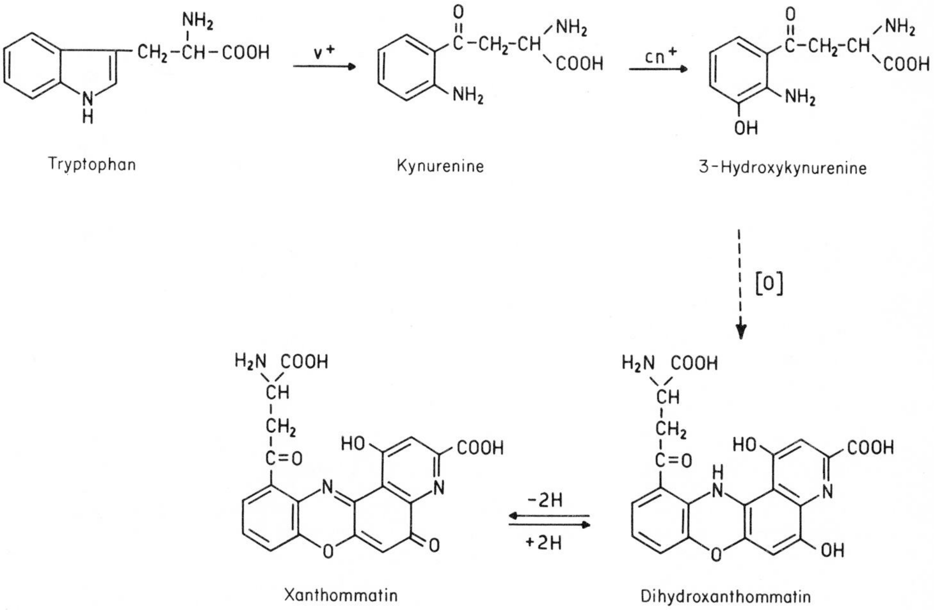

FIG. 3-3. Biosynthetic pathway for ommatin synthesis in *Drosophila.*

PURINES AND PTERIDINES

Among poikilotherms, purines and pteridines are important pigments that differ markedly from one another in a functional sense. Nevertheless, because a close biochemical relationship exists between these two groups of

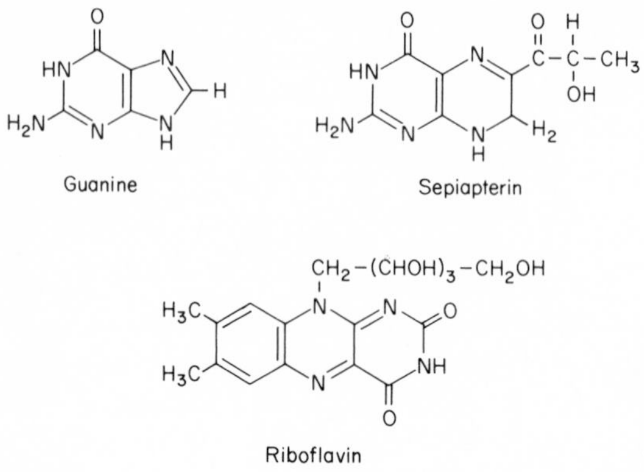

Guanine

Sepiapterin

Riboflavin

FIG. 3-4. Structural relationships between purine, pteridine, and flavin pigments.

compounds (Fig. 3–4), they will be discussed together. Although uric acid, hypoxanthine, and adenine are frequently found as pigments, guanine is perhaps the most widely distributed purine pigment. It was first isolated from guano, from which it derives its name, and late in the 1800s it was shown to be a prevalent pigment of iridophores. Just before the turn of the century, yellowish or whitish pigments were discovered in the wings of certain butterflies and were thought to be related to uric acid. Later these pigmentary substances were named pteridines (*pteros*: wing), and the two primary compounds were called leucopterin and xanthopterin.

Both purines and pteridines contain two rings, one of which is a pyrimidine. In purines, the second ring is a five-membered imidazole group, and, in pteridines, this is replaced by the six-membered pyrazine ring. The biosynthesis of purines is well understood, but much less is known about the synthesis of pteridines although the evidence is rather convincing that pteridines are synthesized through a purine precursor (49). Isotopic labeling experiments indicate that specific positions in the pyrimidine rings of both purines and pteridines have a common origin. The same can be said for analogous positions in the pyrazine and imidazole rings. Moreover, after the incubation of various labeled purines in pteridine-synthesizing tissues, the label can be recovered from the pteridines. In addition, it has been demonstrated recently that oxidation of certain pteridines can lead to a contraction of the pyrazine ring to form purine products (50). Because of the obvious parallelism in the biosynthesis of pteridines and purines, it seems predictable that, in organisms actively synthesizing both types of pigments, some sort of competition for substrates should exist. This seems to be the situation in amphibian larvae wherein conditions favoring the synthesis of pteridines are accompanied by a reduction in purine content and vice versa (51, 52, 53).

During the past few years it has been demonstrated that pteridines play an important role in the bright-colored pigmentation of poikilotherms (23). In large measure, this may be related to the fact that pteridines constitute the only pigmentary component of xanthophores and erythrophores that animals can synthesize. As our understanding of the pigmentary role of pteridines has increased, it has been revealed that a large variety of these compounds exist in nature. Many are of limited distribution, whereas others are ubiquitous. This fact is exemplified most notably by the drosopterins and sepiapterin, which were originally discovered as eye pigments in dipterans and which are now known to constitute an important group of bright pigments of fishes, amphibians, and reptiles. In amphibians, they make their first appearance precisely with the onset of xanthophore differentiation (Fig. 3–5) (24). Apparently pteridine synthesis occurs in the xanthophore, for those cells grown in culture develop the pteridine pattern of the species. Moreover, heteroplastic and xenoplastic transplantation of pigment cell-forming areas results in the appearance of donor xanthophores and donor pteridines on the host (54).

FIG. 3-5. Chromatographic analysis of developmental changes in the pteridine pattern of *Triturus pyrrhogaster* during the course of development. Spots indicate fluorescence of the constituent pigments induced by exposure to ultraviolet light. Stages 1, 2, 3 represent, respectively, neurula, early tailbud, and late tailbud stages. Xanthophores have not yet differentiated and no pteridines are present. The fluorescence in these three stages is of a flavin. Stages 4 and 5 represent early and late larval stages. The xanthophores are fully expressed, as is the complete pteridine pattern. The adult pattern represented by 6 differs markedly from that of the larva. [From (24)]

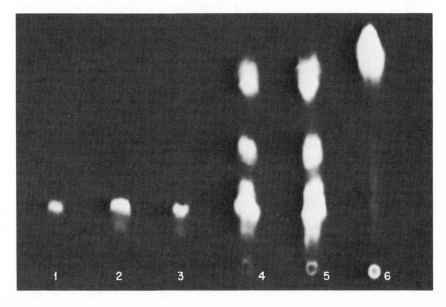

While it is generally thought that pteridines of pigmented tissues exist in an unconjugated form, it is conceivable that colorless pteridines present in high concentration in pigmented skin exist as dimers or in a bound form. It is striking that whenever pteridines are present in the pigmented areas of a species, an array of compounds is found that constitutes the specific pteridine pattern of that species (55). The interrelationships of the various components of the pattern are in need of further study. Presumably, during ontogeny, there is a precise sequential appearance of the various pteridines of the pattern; details of this condition have not yet been demonstrated clearly. The existence of some of the pteridine constitutents of pigmented skin is due to the degradation of other components of the pattern. Pteridines are known to be labile and can be degraded in the presence of light or oxidizing agents. In general, the pteridine pattern does not vary in different parts of the organism. However, deep in the dermis of adult amphibians, a granular, pteridine-containing layer is known to contain relatively higher concentrations of those pteridines that are usually considered to be decomposition products (56).

An important question concerning both purine and pteridine metabolism relates to the occurrence of riboflavin in pigmented tissues. Riboflavin is frequently found in association with the red or yellow pigmentation of amphibians. In fact, large concentrations are found in the bright red or orange pigmentation on the ventral surface of adult newts. Although it has not yet been shown definitely that riboflavin is found in specific cellular organelles in these regions, it is assumed that this flavin is contained in vesicles found in xanthophores or erythrophores of the ventrum. Currently it is believed that riboflavin found in animal tissues is derived from dietary sources; however, its close chemical resemblance to pteridines and purines (Fig. 3–4) leads one to entertain the possibility of a metabolic relationship among the three groups. This relationship is suggested not only because of the known incorporation of the pyrimidine ring of adenine into riboflavin (57) and the fact that riboflavin can be formed from lumazine, a close relative of naturally occurring pteridines, but also because as pteridines are lost from the skin of newts during metamorphosis, riboflavin concentrations increase (55). Again, this area represents a problem in need of investgation.

CAROTENOIDS

Among the most widely distributed of all pigments are a group of yellow, orange, red, or violet compounds found in both plants and animals. Because they occur in such abundance in the carrot, they have been named

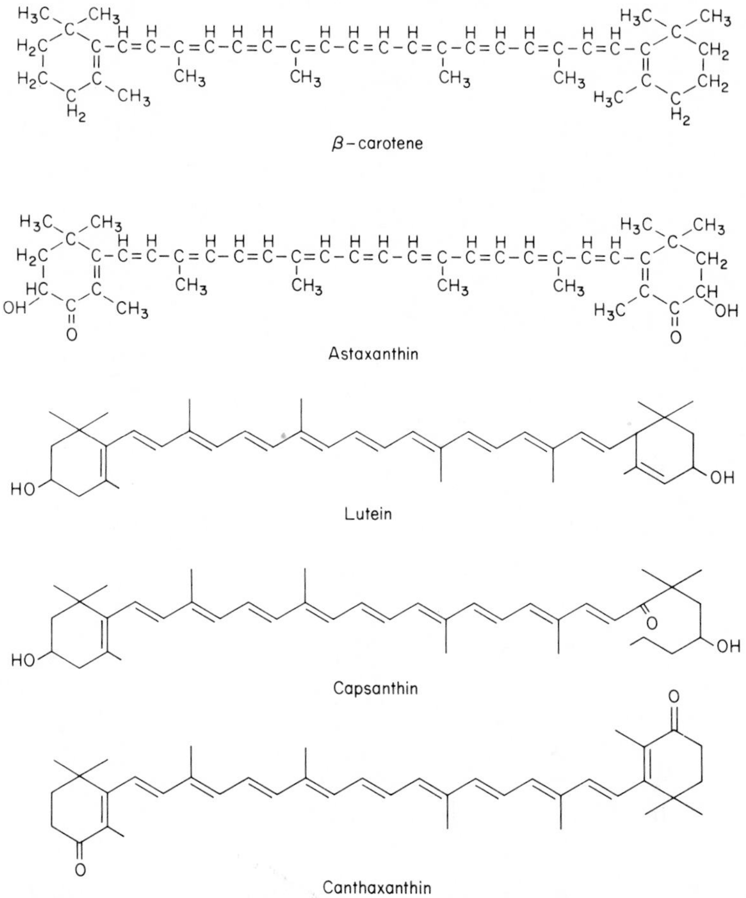

FIG. 3-6. Representative carotenoid pigments. β-carotene is a commonly occurring carotene. Astaxanthin, lutein, capsanthin, and canthaxanthin are commonly occurring xanthophylls.

carotenoids. The general group of carotenoid pigments is divided into two main categories, the carotenes and the xanthophylls (Fig. 3–6). Molecules of both groups consist of four isoprene units that form a chain of carbon atoms linked by alternating single and double bonds (conjugated system). Usually an ionone ring is located at each end of the chain. The number of double bonds in these unsaturated compounds is variable, and it is the degree of unsaturation that determines the specific color characteristics of individual carotenoids. While carotenes are pure hydrocarbons, xanthophylls contain oxygen, and it is this feature that distinguishes the two classes. Although insoluble in water, carotenoids are readily soluble in organic solvents. In nature, they are frequently found dissolved in lipids and thus have been referred to as lipochromes. In earlier times, when it was believed that carotenoids were the primary pigments of yellow, orange, or red pigment cells, these chromatophores were called lipophores because they appear to contain carotenoid pigments in small oil droplets. Carotenoids are not always found in association with lipids; they may be present in the free form and often they exist as carotenoid-protein complexes. In the natural state, such complexes may be purple, blue, green, or brownish in color. For a more detailed treatment of the nature of carotenoids, including nomenclature and biochemical methods, the reader is referred to Fox's book (58).

Although carotenoids are such important pigments among animals, it is interesting that animals cannot synthesize carotenoids and must depend instead on their diet as a source of these compounds. Knowledge of this fact has had practical application for many years as is witnessed by the common practice of feeding captive animals, notably fishes and birds, a high carotenoid diet. Animals have the capacity, however, to modify the carotenoids that are ingested and in this way establish a fairly definitive carotenoid pattern. Recent comprehensive analyses of carotenoid pigments of certain crustaceans have been provided by Gilchrist (59), and the same has been done for feather pigments of genetic variants of the Gouldian finch, *Poephila gouldiae,* by Brush and Seifried (60). The nature of the carotenoid pattern is in part due to a given sequence of metabolic steps that can be studied by feeding organisms pure carotenoids. In this way, Herring (61) was able to show that *Daphnia* is able to form echinenone, canthaxanthin, and astaxanthin, in that order, from β-carotene. Similar findings were made by Hata and Hata (62) from studies on *Artemia*.

The specificity of the carotenoid pattern of an animal is chiefly based on the selectivity of carotenoid uptake, not only general dietary uptake but uptake by tissues as well. Fishes have been the subject studied most in this respect, and it is known that they concentrate carotenoids in the skin, ovary, liver, muscle, and other tissues. Some tissues selectively accumulate only specific carotenoids, a point that is especially true of xanthophores and erythrophores. The specificity of carotenoid transfer in the medaka. *Oryzias latipes* has been investigated by Takeuchi (63) who has shown that while lutein, capsanthin, and canthaxanthin can be transferred readily from the yolk to the larval xanthophores, little carotene and no β–apo–8′–carotenal were transferred. The mechanisms by which tissues select carotenoids are not yet understood; they provide the student of pigmentation with a fertile area for new research.

Our knowledge of the biochemistry of animal pigments has increased markedly during the past decade. In large measure this progress reflects the success of organic chemists who, for example, have made progress in the understanding of eumelanin and phaeomelanin structure and pteridine biosynthesis. Since these investigations are largely concerned with the chemistry of such compounds for the sake of pure knowledge, it behooves us as biologists to be aware of these advances so that they may be applied in the context of modern cell biology.

developmental aspects
of pigmentation

Knowledge of the developmental aspects of pigmentation has accumulated rapidly in recent years. Unfortunately, however, except for the elegant studies of the genetics of eye coloration in *Drosophila,* only fragmentary information is available about the developmental physiology of pigmentation of invertebrates. Vertebrates, on the other hand, have been the frequent subjects of studies dealing with pigment cell development and pigment pattern formation. Therefore, with much regret, our coverage of developmental aspects of pigmentation is limited to vertebrates, particularly to amphibians and mammals.

ORIGIN OF PIGMENT CELLS

A tremendous impetus to the study of pigment cell development of amphibians was the discovery that amphibian chromatophores take their origin from the neural crest (64). The first demonstration of this fact occurred at the time when classical experimental embryology of the Spemann type was beginning to dominate the field of zoology. During this dynamic period, pigment cells or, rather, their anlage, the neural crest, provided a model for the use of classical experimental techniques (65). Amphibians possess special virtues in this respect: embryos are readily available, the neural folds are quite accessible, and pigment cells are distinctive in appearance and chemical composition. Moreover, characteristic differences in pigmentation occur between various species. In some of the early experiments, large areas of the neural folds were extirpated and discarded.

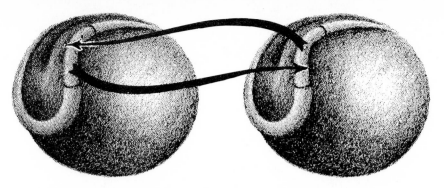

FIG. 4-1. Reciprocal transplantation of neural crest.

Subsequently, it was found that the developing embryo and larva were considerably deficient in the number of pigment cells they contained. These experiments, in conjunction with others involving reciprocal transplantation of neural crest between species with different pigment patterns, provided concrete evidence that amphibian chromatophores take their origin from the neural crest (Figs. 4–1 and 4–2). By using different species that contain a variety of pigment cell types, it was found that all three basic chromatophore types—melanophores, iridophores, and xanthophores—are so derived. These initial conclusions have been confirmed in recent years through the use of *in vitro* techniques in which neural crest cells are cultured in hanging-drop cultures or in explants of epidermis (Fig. 4–3) (66, 67).

FIG. 4-2. Larva of *Notophthalmus viridescens* that had received a neural crest graft (right side) from *Ambystoma mexicanum* at the neurula stage. Dark pigmentation on both front limbs, on the right gill area, and on the dorsal surface between the gills is of donor origin.

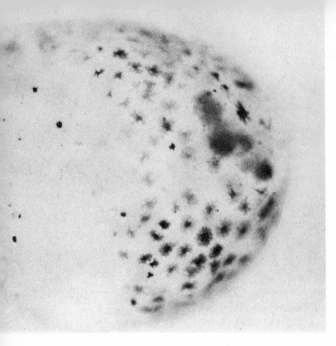

FIG. 4-3. Epidermal vesicle containing melanophores. Formed from the explantation of neural crest with ectoderm from the same embryo.

Although amphibian embryos are amenable to such experimental manipulations, other vertebrates provide greater obstacles. Nevertheless, cogent experimentation has shown that the neural crest is the site of origin of chromatophores of all vertebrate classes. Proof that such is the case for mammals deserves special mention because of the rather interesting experiments that were performed (68). These studies utilized the fact that embryonic mouse tissues transplanted intracoelomically into chick embryos take and remain viable for many days. Therefore, from definitive stages in embryogenesis, neural crest material was transplanted from black mouse embryos into the coelom of white Leghorn embryos of about $2\frac{1}{2}$–days incubation. During the subsequent two weeks, melanoblasts grew out of the crest material to invade mouse epithelial tissues that were included in the graft. Such did not occur in appropriate control transplantations. A fringe benefit from this discovery that has apparently not been commented on before is the fact that these experiments demonstrated for the first time that the epidermal melanophores, in particular, are of neural crest origin. The origin of epidermal melanophores was not determined in the earlier classic experiments on the origin of pigment cells.

MIGRATION OF PIGMENT CELLS

Considering the site of pigment cell origin and the ultimate pigmentation of the integument far from the neural crest region, it is obvious that pigment cells move relatively long distances during normal embryogenesis. This fact has been confirmed by a variety of transplantation experiments, some of which revealed that chromatophores have the capacity to move much farther than they ever do during normal development (69). It has also been shown that chromatoblasts originating on one side of the embryo can cross over to the other side; at least, such is concluded from the appearance and development of chromatophores on the flank of an embryo from which the corresponding neural crest had been totally extirpated. It appears that although the peregrinations of chromatoblasts may be exten-

FIG. 4-4. Three individual *Rana pipiens burnsi:* wild-type chimeric frogs as shown in **A**, **B**, and **C**. A sharp line of demarcation separates the anterior *burnsi* half from the posterior wild type. For comparison, a wild type: wild type chimera is illustrated in **D**. [From (70)]

sive, they occur preferentially in a dorsoventral direction. This point is shown rather clearly in experiments utilizing chimeras formed at the tailbud stage by the grafting of the front half of one species to the back half of another species that has a different pigment pattern. Whether these exchanges are made between species of frogs or between species of salamanders, the line of fusion is always clear and abrupt (70). Relatively few chromatophores appear to cross this point of fusion from one component to the other (Fig. 4–4). It is interesting to note that the limited passage that does occur is from the anterior member to the posterior one. Evidently the majority of pigment cell movements take place before very much pigment synthesis has occurred in the chromatoblast; however, there is evidence that some differentiated melanophores can carry out limited morphogenic movements. Knowledge of the mechanisms of chromatoblast movement is, unfortunately, meager. Nevertheless, observations from tissue culture experiments have contributed some information. In hanging drop cultures, it has been observed that during the first few days, chromatoblasts migrate from the explant by virtue of lobopodia and later by developing filopodia. Of course, there is no way of knowing whether migratory activities on the glass surface of a culture situation duplicate the *in vivo* migration. This is important in view of the fact that it is thought that the nature of the epithelial surface plays a role in chromatoblast migration (71).

PATTERN FORMATION

The results of differential migration of chromatoblasts are evident in pattern formation. For example, the failure of melanoblasts to migrate properly is thought to be the basis for the formation of the well-known white mutant of the Mexican axolotl (*Ambystoma mexicanum*) (71). Similar features of migration have been invoked to explain white spotting patterns in a variety of species, including mice (72, 73). It has been concluded that the absence of melanophores from adult hair follicles results from the failure of neural crest cells to enter the white skin area. Moreover, it is contended that the prevention of melanoblast immigration is attributable to blocking action of the skin itself. Although the latter point has been proven, there is also good evidence that the exclusion of melanoblasts from some areas of the skin is due to factors intrinsic to the neural crest.

The possible role of chromatoblast interaction in determining pigment patterns is an important consideration. Such a possibility is immediately suggested, a priori, by the existence of barred or striped patterns of pigmentation. It has been proposed that although grouping of chromatophores of a given type to form a discrete spot could result from mutual positive affinities, the homogeneous distribution of pigment cells in the interspot area could result from negative affinities. Superimposed on this concept of positive or negative affinities is the possibility that the onset of migration from the neural crest differs with each chromatophore type. Thus melanophores, which presumably migrate first, could dominate a specific area and inhibit the immigration of xanthophores, which are thought to migrate from the crest at a later time. Again, evidence from tissue culture experiments lends credence to such a thesis of chromatoblast interaction, for it has been shown that when isolated groups of individual pigment cells are confined to a small space, the various individual chromatophores will move about to occupy, finally, positions at fixed distances from one another (66). It almost appears as though these cells are conforming to spheres of mutual influence. A possible *in vivo* manifestation of negative chromatophore interaction is demonstrated by transplantation experiments in which neural crest from relatively slow-growing species, such as newts, is grafted to the more vigorous ambystomids. On such combinations the expression of newt pigmentation is suppressed, probably because the outgrowing host chromatoblasts inhibit the migration of those derived from the donor crest (69).

Although the failure of neural crest cells to migrate into a given area of the skin surely deprives that area of pigmentation, the same result can be accomplished by a regional failure of chromatoblasts to survive or even by the inability of neural crest cells to differentiate in the region in question. Even though the former condition seems theoretically possible, it is difficult to discover concrete examples where it occurs. On the other hand, there are numerous examples of the inability of chromatoblasts to differentiate in given regions of the skin. In certain white spotting mutants of mice, and in piebald mice, all indications point to the fact that the melanoblasts migrate into the spotted skin, but apparently they do not differentiate. Transplantation experiments reveal that the same is true for salamanders, and observations made at the level of the electron microscope indicate that while blue mutants of frogs lack yellow pigments, xanthoblasts or semidifferentiated xanthophores are present in their proper location.

In part, pigmentary expression in given areas of the skin has been ascribed to chromogenic activity, either of the immediate chromatophore environment or of some tissue that the chromatoblast comes in contact with during migration. In terms of actual causal relationships, we know little about the basis for chromogenic activity, although it appears that in some cases it might relate to the availability of substrates necessary for pigment synthesis. That there are regional differences in the chromogenic nature of the embryonic epidermis is evident from experiments on salamanders involving reciprocal transplantation of dorsal, ventral, and flank epidermis. For example, the transplantation of dorsal epidermis taken from crestless embryos to

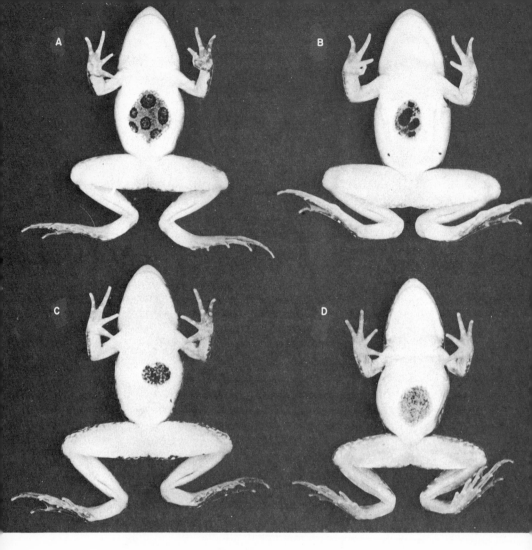

FIG. 4-5. Ventral surfaces of four young adult *Rana pipiens burnsi* showing the expression of dorsal pigmentation resulting from the transplantation of neural crest to the belly region at the neurula stage. Different mutants are used in the transplantation The pigmentation that differentiates in the transplant area is characteristic of the donor. [From (74)]

the belly surface of intact hosts results in the occurrence of an island of dorsal pigment pattern on the ventrum. In contradistinction to these experiments, others champion, with good reason, the role of the neural crest in the differentiation of pigment patterns. Presumably, one of the best examples results from the transplantation of neural crest from spotted wild-type *Rana pipiens* embryos to the belly of the unspotted *R. pipiens burnsi* mutant variety (74). A discrete and clearly delineated island of spotted dorsal-type integument develops on the belly of the *burnsi* host after metamorphosis (Fig. 4–5). It is attractive to interpret this finding as an indication of the autonomous role of the neural crest in determining pigment pattern; however, considering the known role of the epidermis in determining pigment pattern, it seems wisest to reserve judgment at this time. The principal reason is that most investigators who work with the neural crest transplant the entire fold and, consequently, include epidermis and other elements in their graft. Therefore, the influence of epidermis and other neural derivatives in determining the nature of the graft pattern cannot be excluded. Because of these considations, it seems that the question of whether the neural crest operates autonomously in determining pattern is not resolved.

On the positive side, there is good evidence that pigment cell differentiation is an autonomous function of the neural crest cell. It has long been known from reciprocal xenoplastic neural crest transplants that donor-type chromatophores appear on the host. Often these chromatophores are widely dispersed, thus excluding the possible involvement of donor ectoderm as suggested above. The discovery that amphibian xanthophores have a characteristic pteridine pattern has been important in this regard, for it has been demonstrated *in vitro* that as xanthophores develop in neural crest transplants, the full pteridine pattern appears (24). Moreover, when reciprocal neural crest transplantations are made between species having distinctively different pteridine patterns, the characteristic donor pteridines are found in various areas of the host (54).

DETERMINATION OF CHROMATOPHORE TYPES

That the neural crest can give rise to so many different kinds of cells, including at least the four basic types of chromatophores, is one of the great unsolved problems of pigment cell development. Some important questions are as follows: Are all neural crest elements determined at the same time? Is there a sequence in the determination of pigment cells? What is the mechanism of this determination? Is there an interconversion between various chromatophore types? At present these questions cannot be answered; fortunately, however, there is a positive side to the picture. Although we do not know, for example, when chromatophore determination occurs, there is good evidence that it happens very early, even before the neural folds appear. It is known that there is a definite sequence in the appearance of the various chromatophore types: differentiation of dermal melanophores occurs first, followed by xanthophore differentiation, and, finally, iridophores are expressed. Epidermal melanophores make a relatively late appearance; for

example, in *Xenopus,* the South African clawed toad, epidermal melanophores do not appear until shortly before metamorphosis.

With respect to the interconversion of the various pigment cell types,
supporting evidence exists, which is, however, circumstantial. For example,
xanthic goldfish can be melanized by hormone treatment (76, 77). Moreover, their pterinosomes exhibit tyrosinase activity that is normally characteristic of melanosomes (78). Furthermore, chromatophores of one type
sometimes contain additional pigmentary organelles (Fig. 4–6) characteristic
of other types (29, 79, 80). One of the best examples is to be found in the
dove iris, in which it was observed that iridophores contain reflecting
platelets and melanosomes (19). Even more striking is the fact that some
of the reflecting platelets appear to be partially melanized (Fig. 4–7), or,
as noted in cells of the tapetum lucidum of the teleost *Dasyatis sabina,* both
melanosomes and reflecting platelets may occur within the same surrounding
membrane of the formative vesicle (81). Genetic information also supports
the concept of chromatophore metaplasia in that melanoid mutants occur
frequently among amphibian species. The melanoid condition is expressed
by an increase in melanophore numbers accompanied by a corresponding
decrease in the number of xanthophores and iridophores (82). It has been
suggested that such genes operate at the level of the neural crest to switch

FIG. 4-6. Xanthophore from the dorsal surface of an adult *Anolis carolinensis* containing a few melanosomes interspersed between the many pterinosomes that are
typical of this chromatophore. ×16,000. [From (79)]

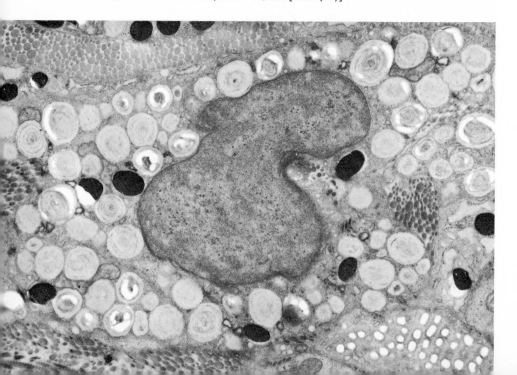

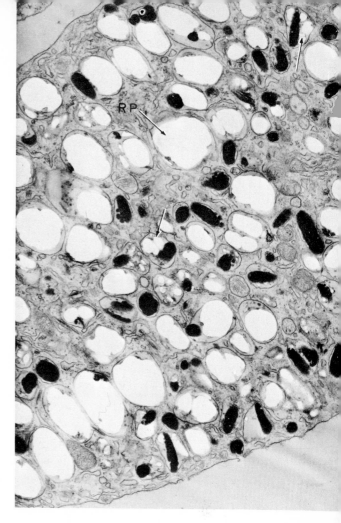

FIG. 4-7. Iridophore in the iris of the Inca dove containing typical reflecting platelets (RP) and reflecting platelets that are partially melanized (arrows). [From (19) and (80)]

chromatoblast differentiation in the direction of melanoblast formation. Other evidence is derived from hormonal effects on chromatophores, which will be described subsequently.

Ultimately, of course, the determination of chromatophore type and the expression of chromatophore differentiation are based on gene action. Thus the accumulation of pertinent genetic information is an important goal in understanding the development of pigmentation. We are fortunate that among various species, both vertebrate and invertebrate, a number of well-described pigmentary mutants exist. Despite the fact that the mechanism of action of most of these genes is not thoroughly understood, genetic information has been a valuable research tool. In the previous paragraphs, the use of various pigmentary mutants has been alluded to and still many other examples exist that are not in the immediate context of our discussion. As for the future, it seems probable that the use of various genetic strains of mice will be exceedingly helpful, for many different pigmentary mutants

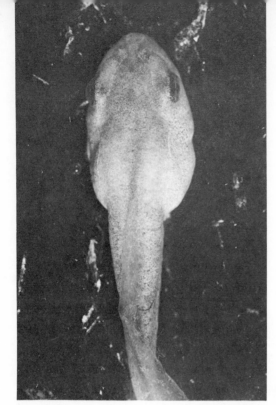

FIG. 4-8. Silvery tadpole of *Rana sylvatica* resulting from the removal of the hypophysial primordium from a tailbud embryo.

A

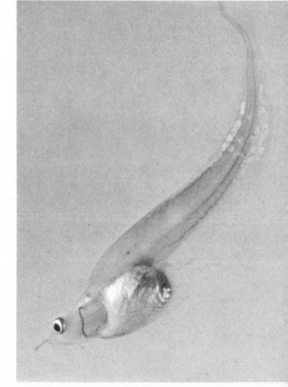

FIG. 4-9. Larvae of *Xenopus* laevis. **A**, hypophysioprivic; **B**, intact. Note the absence of dark pigmentation and the presence of iridophores in the tail fin of the hypophysioprivic larva.

are available, and their genetics are already understood (83). A recent example of the use of such mutants is derived from experiments in which two mouse embryos of different genotypes were fused at cleavage to form a mosaic embryo (84). The standard pigment pattern of mice produced in this way consists of 17 bands of alternating color on each side of the animal. It has been suggested that these bands represent clones of melanophores that are derived from 17 primordial melanoblasts. The significance of this finding needs clarification.

HORMONAL EFFECTS ON CHROMATOPHORE DEVELOPMENT

Since the early days of endocrine investigation, it has been known that hormones exert a profound influence on the coloration of animals. So striking were the pigmentary effects resulting from ablation of the hypophysial placode (anlage of the adenohypophysis) of anurans that the silvery tadpoles so obtained were described as being albinistic (85) (Fig. 4–8). We now

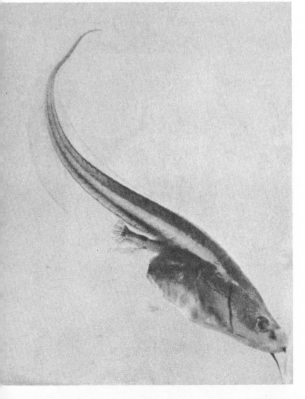

B

know that what was observed here was a profound manifestation of morphological color change. In contrast to those of normal larvae, melanophores of hypophysioprivic larvae are consistently reduced in number and size and are in a punctate state. Iridophores seem to be increased in number and size and their pigment is maximally dispersed. This situation can be reversed by the administration of intermedin preparations. It should be emphasized at this point that these hormonal effects are at the level of chromatophore expression and not at the level of some pattern-determining mechanism. This point can be shown most clearly by causing hypophysioprivic larvae to metamorphose by thyroxine administration. In such cases, the newly transformed froglets display a ghostlike, but exact, manifestation of the usual spot pattern. We know of no cases wherein the action of a hormone determines pattern directly; however, indirect effects on color pattern cannot be excluded. A clear example is shown in the case of hypophysioprivic *Xenopus* tadpoles (Fig. 4–9). Intact larvae of these species do not display iridophores on either the dorsal surface or the tail fin. However, in the absence of the hypophysis, there is a striking manifestation of iridophores in both places. This response should probably be categorized technically as an extreme case of morphological color change with the consideration that, in intact larvae, iridoblasts occupy these positions but are incapable of expressing themselves because of the inhibiting action on pigment synthesis exerted by intermedin from the hypophysis (Fig. 4–10) (86). Consequently, pigment cells become visible in places where they are normally not seen. The pigmentary pattern can be affected by chromatophore proliferation. Convincing evidence is presented by Pehlemann, who shows that prolonged stimulation by intermedin results in a marked increase in mitotic activity of existing melanophores (Fig. 4–11) (87). In the final analysis, it seems reasonable

FIG. 4-10. *Pleurodeles waltli* larvae. Normal, left; hypophysioprivic, right.

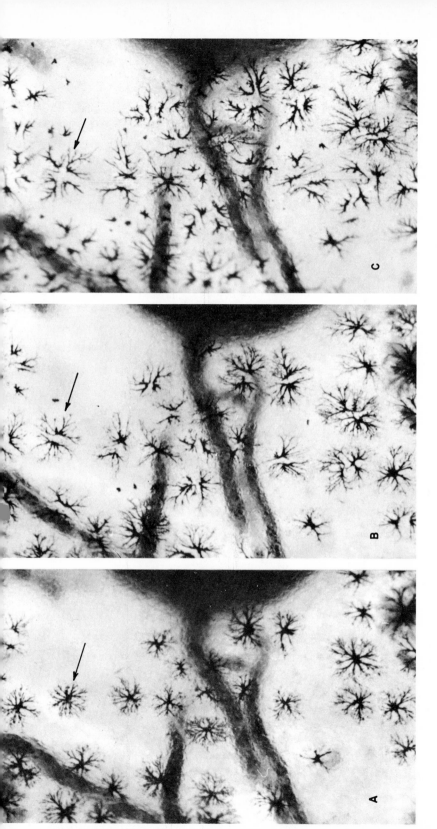

FIG. 4-11. Detail of the upper right side of the head of a *Xenopus* larva. (**A**) 1, (**B**) 5, (**C**) 7 days after transection of the hypophysial stalk. A possible increase in circulating level of chromatophore-stimulating hormone may be responsible for the resulting mitoses of melanophores. A single melanophore (arrow) can be traced through two mitotic divisions in **B** and **C**. [Courtesy of Dr. F. W. Pehlemann (87)]

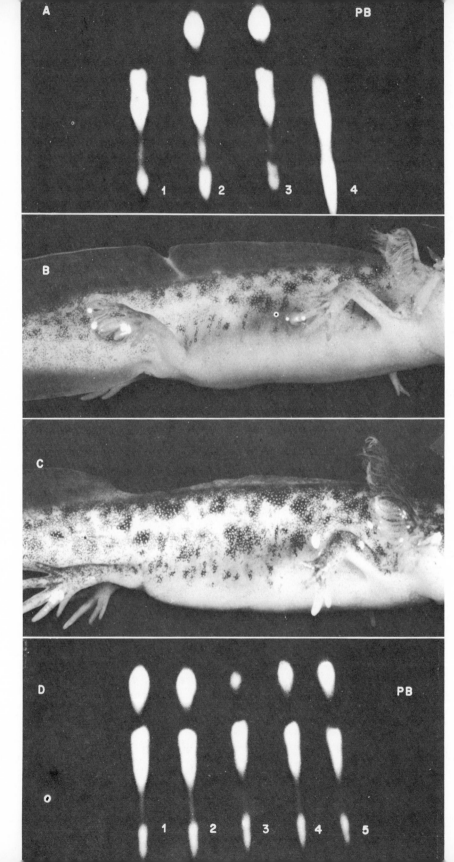

to consider that hormones can affect chromatophore development through the mechanisms of morphological color change.

Hormones other than those of hypophysial origin act on chromatophore development. Thyroxine is one of the best examples. In metamorphosing amphibians, expression of the intrinsic color pattern of the animal is caused by the direct action of thyroxine. It is manifested by a profound increase in the number of melanophores, iridophores, and xanthophores. In teleost fishes, the thyroid also appears to affect pigment development. For example, accompanying parr-smolt transformations in salmonids, there is apparently an increase in guanine and a decrease in melanin. Whether these changes reflect a direct action of thyroxine on pigmentation or an indirect action of the hormone affecting intermedin release is not immediately clear. One of the most specific effects of thyroxine on pigmentation is demonstrated by the change in pteridine pattern that occurs at metamorphosis in newts (88). Typical of the larval pteridine pattern of newt xanthophores is the presence of a compound called pleurodeles-blue, which, in normal development, disappears abruptly from the skin at metamorphic climax (Fig. 4–12 A, B, C, D). That the specific disappearance of this substance is mediated directly by thyroxine was demonstrated by experiments in which local metamorphosis was induced by the implantation of thyroxine-containing pellets. Local metamorphosis was accompanied by a localized diminution of pleurodeles-blue.

It should be evident from the points discussed in this chapter that a considerable amount of descriptive information is available about the development of pigment cells and the establishment of pigmentary patterns. Unfortunately, much less is known about the many mechanisms involved. Clearly, certain developmental cues operate on the pigmentary system during ontogeny just as they do on any developing system. The search for knowledge of the translation of these cues into specific developmental events is an important challenge that needs to be accepted.

FIG. 4-12. A. Development of pteridine pattern in skin of *Pleurodeles waltli* larvae 1. young larva; 2. half-grown larva. Note the appearance of *Pleurodeles* blue (PB), a pteridine specific for the newt family; 3. older larva just prior to metamorphosis; 4. newly metamorphosed larva. Note absence of *Pleurodeles* blue. **B.** Control *Pleurodeles* larva. Implanted pellet (arrow) containing only cholesterol. Notch in tail fin is a wound resulting from implantation of the pellet. **C.** Experimental larva in which local metamorphosis has been achieved by the implantation of a pellet containing 20 percent thyroxine–80 percent cholesterol. Note the marked reduction in the height of the tail fin at the level of the hind limbs. **D.** Chromatographic analysis of skin squashes of the animal shown in **C.** 1. tail region; 2. immediately behind the implant; 3. at the site of implantation where local metamorphosis has occurred; 4. in front of implant; 5. head region. The chromatogram reveals the localized disappearance of *Pleurodeles* blue, coinciding with induced local metamorphosis.

chromatophore control I

pituitary role in the control of vertebrate chromatophore responses

The ability of animals to change color either rapidly or more slowly over longer periods of time involves the movement and synthesis of pigmentary organelles within integumental chromatophores. The factors affecting pigment cell responses within and between species are often difficult to assess. The reason is that chromatophores may be affected by either hormonal or neurohumoral agents as well as by direct environmental influences. In addition, a number of endocrine glands release hormones that may affect pigment cells directly or indirectly by inhibiting the release of yet another hormone influential in chromatophore regulation. Certain hormones may be inhibited from being released under conditions of illumination, whereas others may be released under conditions of darkness, or vice versa. In either situation the chromatic responses may appear similar, although their regulatory basis may be quite different.

Chromatophores may also be directly sensitive to such environmental factors as illumination or temperature, or both. To complicate matters further, both the direct and indirect effects of hormones and other stimuli may be proceeding at the same time, and certain chromatophores may exhibit differential responses to such stimuli. Some chromatophores appear to undergo developmental changes with respect to their sensitivity to hormonal or other stimulation. Even the regulation of release of hormones may possibly change with developmental age.

CONTROL OF VERTEBRATE
CHROMATOPHORE RESPONSES

The early work of Smith (85, 89) and Allen (90) clearly established that the pituitary gland was in some way involved in the control of integumental coloration in amphibians. After a number of important investigations, Hogben and Winton (91) were able to conclude that color changes in amphibians result from the action of an agent released from the pituitary gland. The discovery by Swingle (92) that the implantation of an isolated frog intermediate lobe into hypophysectomized tadpoles causes darkening was strong evidence for its being the source of intermedin (melanophore-stimulating hormone, MSH), the chromatophore-stimulating hormone. In amphibians, this chromatophorotropic hormone disperses melanosomes within melanophores and also causes reflecting platelet aggregation within iridophores and dispersion of pigmentary organelles within some xanthophores.

STRUCTURE AND BIOASSAY
OF CHROMATOPHORE-
STIMULATING HORMONES

A number of MSH peptides have been isolated from mammalian pictuitaries (Fig. 5-1) (93). Common to all these peptides is the heptapeptide sequence: Met·Glu·His·Phe·Arg·Try·Gly, which itself is a potent pigment granule mobilizing agent. The presence of the heptapeptide structure, as well as the entire amino acid sequence of α-MSH within the corticotropin (adrenal corticotropic hormone, ACTH) molecule, undoubtably explains why ACTH has chromatophore-stimulating activity. Within an individual sheep pituitary there may be present two different corticotropins, two different β-MSHs, α-MSH, α_2-CRF, and β-lipotropin—seven molecules in all, and all possessing related amino acid sequences and melanocyte-stimulating activity (94).

Although the structures of a number of mammalian MSH peptides are known, the only other vertebrate chromatophorotropin to be structurally characterized is that of an elasmobranch (95). Extractions of 4000 neurointermediate lobes from *Squalus acanthias* yielded two very similar peptides, both closely resembling α-MSH.

Peptide I: H·ser·met·glu·his·phe·arg·try·gly·lys·pro·met·OH
Peptide II: H·ser·met·glu·his·phe·arg·try·gly·lys·pro·met·NH$_2$

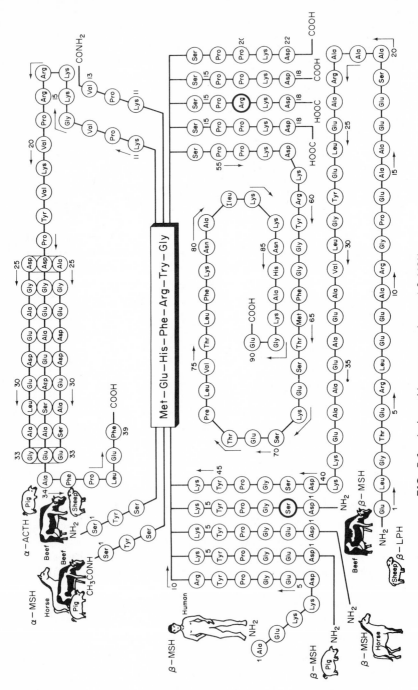

FIG. 5-1. Amino acid sequences of α- and β-MSH, and their relation to corticotropin and β-lipotropin. [From (93)]

These peptides were present in approximately similar amounts and possessed identical melanosome-dispersing activities. A third peptide, found in smaller amounts, was even more similar to α-MSH in that a tyrosyl group was also present. This dogfish MSH, however, is deficient in an acetylated seryl-tyrosyl sequence and has methionine, rather than valine, substituted at the C-terminus (96). No trace of β-MSH was found. Nothing is known as to the structure of MSH or related peptides of other vertebrates other than some preliminary observations of their electrophoretic mobilities (97).

The most sophisticated assay for MSH or other related chromatophore-stimulating hormones or agents is the *in vitro* photometric reflectance method. Although originally devised for use with frog skins (35), this assay can be modified for use with integuments of other vertebrates, such as the lizard. In this assay, the movement of pigment organelles within dermal chromatophores in response to hormonal stimulation results in a lightening or darkening of skins as measured by reflectance changes from the outer (epidermal) surface. In the frog, reflectance changes result from both melanosome movements within melanophores and reflecting platelet movements within iridophores. On the other hand, in the lizard, *Anolis carolinensis,* melanosome movements account entirely for reflectance changes from the skin (37). The *in vitro* chromatophore responses of both the frog and *Anolis* to MSH are similar to those that normally take place in intact animals.

The fine sensitivity of the bioassay is evidenced by the fact that frog (*Rana pipiens*) skins will show a measurable darkening to a 10^{-11} M concentration of MSH (98); lizard (*Anolis carolinensis*) skins are almost as sensitive (99). Modifications of this assay allow measurements of MSH when only small quantities are available (100). The sensitivity of the assay may well depend on both the species of frog from which the skin is obtained and the source of the intermedin. One can obtain a clear dose-response relationship to MSH by using either *Rana pipiens* or *Anolis carolinensis* skins, and the response of individual skins to MSH is uniform. Skins darkened by MSH subsequently relighten when rinsed in fresh Ringer solution (Fig. 5–2).

The *in vitro* frog skin bioassay can be used to demonstrate effectively the chromatophore-stimulating potencies of the amphibian pituitary and to delineate further the darkening potencies of both the pars distalis and the pars intermedia. The pituitary gland of the frog is clearly divisible into its three major lobes. The adenohypophysis can be separated from the pars nervosa, and then the pars intermedia can be cleanly and completely separated from the pars distalis. The pars distalis is many times larger than the pars intermedia; but when bioassayed at similar dilutions, pars intermedia homogenates of either the frog or the rat are much more effective in darkening frog or lizard skins than are pars distalis homogenates. As little as 1/4000 of a frog intermediate lobe can be detected by these bioassay

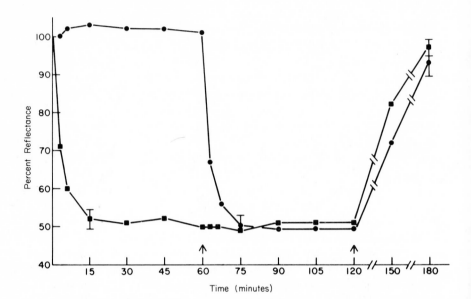

FIG. 5-2. Demonstration *in vitro* of the uniformity of responses of *Anolis* skin to MSH. One group (■) of skins was darkened with MSH for 60 minutes and then the Ringer control group (●) was subjected to a similar concentration of MSH. At 120 minutes, the skins in each group were rinsed in Ringer solution several times. Each point on the graph is the mean of eight measurements of reflectance. Vertical lines represent the standard error of the mean. [From (99)]

methods. Although little is known about the structural features of amphibian chromatophorotropins or corticotropins, there is obviously either a qualitative or a quantitative basis for the differences between the MSH-like potencies of the pars distalis and the pars intermedia. It is possible that the darkening response of skins to pars distalis homogenates could imply a contamination by the pars intermedia, but these results clearly rule out a contamination of pars intermedia homogenates by the pars distalis. Neurohypophysial hormones, as possible contaminants of pars intermedia homogenates, do not themselves darken frog or lizard skins or inhibit darkening by other agents.

Other *in vitro* and *in vivo* (1) bioassays for MSH are in use, as described earlier, and involve subjective estimates of melanophore responses utilizing the Hogben and Slome (34) index (Fig. 2–20) in the determination of the degree of melanosome aggregation or dispersion. Some workers have utilized similar methods to observe iridophore responses of amphibians, plus erythrophore and xanthophore responses of fishes, and to determine the effects of intermedin or other hormonal or pharmacological agents.

PARS INTERMEDIA (MSH)
REGULATION OF MELANOPHORES

Pars intermedia ablation (101, 102) in amphibians results in failure of chromatophore control, thus clearly implicating a role for the intermediate lobe and MSH in color regulation. Prevention of pars intermedia development (103) results in silvery-colored larvae which develop into light-colored adult frogs that are unable to adapt to dark-colored backgrounds (Fig. 5–3). Since the pars distalis and pars nervosa apparently develop normally in these experiments, this factor might rule out a role for these pituitary components and their secretory products in chromatophore control.

Intermedin is undoubtedly the pituitary hormone mainly responsible for regulating vertebrate color changes, but the excessive release of corticotropin in certain disease states may be responsible, at least in humans, for some hyperpigmentary conditions. Adrenal cortical steroids apparently influence vertebrate pigmentation mainly through a feedback control of the hypothalamic regulation of pituitary ACTH release, rather than by any direct effects on melanophores. There is evidence that β–MSH secretion may be regulated by the same factors that control ACTH release. In primary adrenal insufficiency (Addison's disease), there is a greatly lowered level of circulating glucocorticoids, which results in an increased release of ACTH

FIG. 5-3. A. Pale, recently metamorphosed frog, *Hyla regilla*, previously hypophysectomized in the larval state. Pars intermedia is absent, but a regenerated pars distalis is present. **B.** Normal (control) frog of the same age as the pale one. [From (103)]

A   B

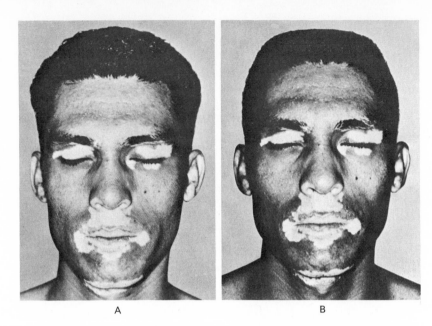

A   B

FIG. 5-4. Effect of α-MSH on patient with vitiligo. **A.** Before hormone treatment **B.** After 23 weeks of hormone treatment. Darkening of nonvitiligous areas of the skin is evident. [From (3)]

followed by hyperpigmentation. The increased melanin pigmentation may then result from melanophore stimulation by ACTH. Because of an increased level of circulating MSH in the blood of humans with Addison's disease (104), it is difficult to relate the hyperpigmentary effects directly to either ACTH or MSH. However, it has been demonstrated (105) that either ACTH (in large concentrations) or MSH injected into human subjects causes increased darkening of the skin (Fig. 5–4).

In Cushing's disease, high circulating levels of ACTH may also be a contributing factor to hyperpigmentation. It is also possible, however, that the increased plasma cortisol (hydrocortisone) levels characteristic of this condition may suppress the secretion of MSH, resulting, then, in a depigmentation that has been noted in a few Negroid individuals with Cushing's disease (106).

In the deer mouse, *Peromyscus maniculatus,* adrenalectomy is followed by darkening of the hair within one to three months (107). Again, this hyperpigmentary state may result from increased circulating levels of ACTH or MSH, or both. It was suggested (107) that "Since the functional level of the adrenal cortex in rodents is related to population density and a host of other environmental factors, differences in pigmentation patterns between populations and among individuals within populations could, in some cases, be due to stress-MSH effects." The suggestion that MSH may indeed regulate coat color in mice is supported by the finding (100) that a transplantable pituitary tumor caused an increase in hair pigmentation. The demonstration that tumor transplants into hypophysectomized mice produced darkening without stimulating the adrenals can be considered good evidence that the hormone being secreted was probably MSH rather than ACTH. In certain genotypes of mice, injections of MSH cause a profound darkening of newly grown hair (108) (Fig. 5–5). In the guinea pig (109), injections of MSH increase the proportion of dark- to light-colored hairs. This effect is greater in black animals than in red ones. These observations, and others, indicate that follicular melanophores are responsive to MSH just as are interfollicular melanophores of man. The genetic background is apparently important in mammals in determining whether an individual animal will respond to MSH or not. These examples of morphological color changes involve alterations in pigment synthesis and are slow compared to the physiological color changes of poikilotherms, which involve rapid pigment redistribution within chromatophores.

Other observations implicating a natural role for MSH in the normal regulation of color in pelage cycles of mammals are those on the short-tailed weasel, *Mustela erminea bangsi* (110). Weasels hypophysectomized in both the winter (white) or summer (brown) coats molt to or maintain the white coat color. These hypophysectomized animals treated with either ACTH or MSH regrow pigmented hair typical of the summer pelage after hair growth is initiated by plucking fur. Hypophysectomized uninjected controls regrow white hairs after hair growth is similarly induced. Other pituitary hormones, such as follicle-stimulating hormone, luteinizing hormone, and thyroid-stimulating hormone, as well as cortisol, have no effect on pelage color in hypophysectomized weasels. Intermedin and ACTH probably exert

FIG. 5-5. Effect of injected α–MSH on hair coloration of yellow mice following plucking. Black areas correspond to the areas of new hair growth. (Courtesy of Dr. I. I. Geschwind)

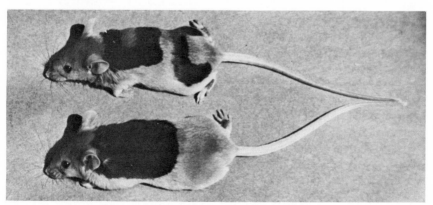

their effects directly on melanophores of hair follicles rather than indirectly via the adrenal steroids.

These effects of ACTH and MSH on skin and hair color of man and mammals are examples of morphological color changes and result from the synthesis of melanin by epidermal melanophores. Hormones may influence melanogenesis within mammalian melanophores by initial actions on the intracellular distribution of melanosomes. The fairly rapid (24 hours) darkening of human subjects treated with MSH suggests that the dispersion of melanosomes within melanophores may precede the onset of new melanin formation (111). There is cytological evidence regarding the action of certain hormones on melanophore morphology, melanin synthesis, and the release of melanosomes to surrounding epidermal cells (3). Mammalian epidermal melanophores may react to hormones in a manner similar to that of the dermal melanophores of amphibians. Actually, it has been demonstrated, for the amphibian at least, that MSH causes a rapid dispersal of melanosomes within epidermal melanophores. The change in intracellular distribution of melanosomes within epidermal melanophores results in a color change that is demonstrable by quantitative estimates of light reflectance and transmission as well as by direct visual comparisons (6). The epidermal melanophores of amphibians and mammals are comparable both in their morphology and in their capacity to provide epidermal cells with melanosomes (Fig. 2–2). However, the capacity for rapidly reversible intracellular translocations of melanosomes so readily demonstrated in the epidermal melanophores of amphibians appears to be less well developed in mammalian melanophores.

With respect to MSH, therefore, the response of epidermal melanophores is quite similar to that of dermal melanophores. The larger size of the dermal melanophores, however, is responsible for their greater importance in physiological color changes of poikilothermic vertebrates. On the other hand, epidermal melanophores of amphibians, as well as those of reptiles, birds, and mammals, including man, play important roles in morphological color changes (40). Indeed, it is, in part, the differences in melanin synthesis within human epidermal melanophores that provides the easily observable racial differences in skin pigmentation. Within any one race of humans, seasonal changes in melanin synthesis within epidermal melanophores as affected by solar radiation provide the means for alterations in skin color, as takes place during sun tanning. Whether MSH (or ACTH) plays a role in these racial or intraracial color differences is not known, but studies with other mammals now make such a suggestion a possibility.

In tissue culture, melanogenesis in xanthic goldfish (*Carassius auratus*) skin is directly correlated with ACTH concentration and not with MSH concentration (112). A pituitary principle is responsible for melanogenesis, for stress elicits melanophore formation in intact goldfish but not in hypophysectomized fish. In addition, the implantation of pituitary material into hypophysectomized fish results in the formation of melanin, which is not induced by similar implantations of other tissues. Injections of ACTH stimulate melanogenesis in both intact and hypophysectomized fish, whereas MSH has no such action. Surgical removal of the pars intermedia, which contains MSH-like activity, does not affect the melanogenesis occurring after stress. If the pars tuberalis and pars distalis are removed, however, melanogenesis fails to occur. These results implicate a role for ACTH rather than MSH in morphological color change in the xanthic goldfish. The melanophores in the scales of the gray variety of goldfish do respond to MSH, however, and it is possible, therefore, that MSH may play some role in the rapid chromatic responses to background adaptation in the pigmented goldfish. In the teleost *Fundulus heteroclitus,* ACTH has no melanogenic effect, but injections of prolactin stimulate melanogenesis in already existing melanophores (113). Intermedin stimulates melanophore proliferation, and these effects are potentiated by prolaction (114).

At present, no unifying generalization can be made as to the possible role of MSH or ACTH in the normal regulation of either physiological or morphological color changes in teleosts. Some fish respond to MSH or other pituitary preparations by becoming darker in color, whereas others become lighter (115). Parker has probably correctly surmised (1), that "It appears that attention must be given to the diversity of pituitary extracts, the production of which is far from uniform, and to possible differences in the chromatophores themselves even in the same fishes." Differences in melanophore responses to MSH or other agents may well depend on whether the pigment cells are innervated or not. The stress induced by merely handling fish while making injections may be more responsible for the ensuing color changes than are the injected materials themselves. Melanosomes within denervated melanophores of *Fundulus* disperse in response to a pituitary extract from other such fish (116). This fact indicates that melanophores of this fish can respond when released from nervous control. Even after hypophysectomy, however, some fish (e.g., *Fundulus*) can rapidly adapt to a black background. Although the catfish, *Ameiurus nebulosus,* can still darken after hypophysectomy, it will not darken to the same extent as a normal fish.

These results indicate that melanosome dispersion in some fish is mediated not only by MSH but possibly also through "dispersing nerves" (115) innervating the melanophores. In rapid adaptations to dark-colored backgrounds, these nerves are effective before the pituitary secretion comes into play, but this neurogenic stimulus is less effective than is MSH. Considering the mobility of fish and the likelihood, therefore, of frequently encountering new backgrounds, it is possible that nervous regulation of melanosome movements would be of more immediate importance in melanophore regulation than would be the secretion of a relatively slower acting MSH. Under conditions of prolonged maintenance over a dark-colored background, however, the presence of circulating MSH might enhance the adaptive response. The fact that most amphibians are much less mobile than fish and do not normally encounter changing background conditions that would necessitate rapid color changes might explain the apparent absence of direct nervous regulation of chromatophores in this vertebrate group. Considering the early evolution of and great diversity of teleosts, it could be expected that the relative importance of either nervous or humoral regulation of chromatic responses within this large vertebrate group would be highly variable.

Although only a few studies on reptilian chromatophore control have been undertaken, it is clearly established that the hypophysis plays a role in chromatic regulation in those few reptiles that have been studied. Intermediate lobes from either the frog or from *Anolis* itself, when injected into hypophysectomized or intact *Anolis,* result in a rapid darkening of the skin (117). This darkening is complete in only a few minutes, and the time course of the darkening is similar to that normally required for the lizard to change from green to brown. Isolated pieces of *Anolis* skin also darken when placed in solutions containing extracts of intermediate lobes (118) or MSH (119).

Hypophysectomy of the lizard, *Hemidactylus brookii* (120), and the rattlesnake, *Crotalus viridis* (121), results in permanent pallor. This pallor also results if only the intermediate lobe is removed. Injections of MSH into the rattlesnake (121) results in a darkening of the skin that can be duplicated *in vitro*. This darkening results from melanosome dispersion within both dermal and epidermal melanophores. Injections of either MSH or ACTH into *Chameleo jacksoni* darkens this lizard (122). These results clearly suggest that the pars intermedia and its release of intermedin, as in amphibians, regulate melanophore responses in some species of reptiles. The relative importance of pars intermedia secretions in chromatophore control among different species of reptiles, however, is yet to be established. Certainly, as will be discussed later, chromatophores of chameleons are normally controlled through nervous innervation.

Both adult (metamorphosed) and larval cyclostomes (*Lampetra,* 123; *Mordacia* and *Geotria,* 124; *Myxine,* 125) remain dark on either a white or a black background. Both larvae and adults blanch after hypophysectomy because of melanophore contraction. Injections of mammalian pituitary extracts into hypophysectomized individuals result in darkening of the animals. Loss of the meta-adenohypophysis (where intermedin is localized) is mainly responsible for loss of chromatic control (123, 126). This is good evidence that MSH is probably the pituitary agent responsible for melano-some dispersion in cyclostomes.

Melanosomes within melanophores of some elasmobranch fishes ag-gregate and disperse in response to adaptations to a white- or black-colored background. There is strong evidence within this vertebrate group that these responses are regulated by MSH (1). Hypophysectomy results in a per-manent pallor in sharks, but these fish will darken in response to injected pituitary extracts. Injections of blood from dark dogfishes darken pale indi-viduals. Similar injections into darkened individuals have no observable effect (127). No color changes are elicited when blood from pale dogfishes is injected into either pale or dark individuals. Loss of the neurointermediate lobe in elasmobranchs is most specifically related to pallor following removal of the various parts of the pituitary.

Only with the vertebrate class Aves, the birds, has there generally been a failure to implicate intermedin in the regulation of integumental and feather coloration. It is interesting, therefore, that the avian pituitary gland lacks an anatomically distinct pars intermedia (128). The pituitary gland of the Rhode Island Red chicken was reported, however, to contain large amounts of MSH (129). In the weaver finch, *Stegnanura paradisaea*, continuous in-jections of MSH or ACTH both fail to melanize the feathers, confirming an earlier negative result using MSH (130). However, it has wisely been sug-gested that a final exclusion of intermedin may require the use of avian MSH (131), if such exists. There is some evidence that the development of black down feathering in hybrid chick embryos may be experimentally in-fluenced by ACTH and α-MSH (132).

INTERMEDIN REGULATION OF
BRIGHT-COLORED CHROMATOPHORES

The foregoing discussion of the role of MSH in the regulation of vertebrate integumental pigmentation has mainly dealt with melanophore regulation. Other cells, bright-colored pigment cells, are present within the integument of most poikilothermic vertebrates and play a role in color changes. Birds and mammals lack such integumental bright-colored pigment cells, and melanophores are apparently the only chromatophores possessed by cyclostomes (1).

Considering the common neural origin of all vertebrate chromato-phores (133), it is not surprising that the bright-colored pigment cells—the xanthophores, erythrophores, and iridophores—might be expected to respond to similar endocrine stimulation as do the melanophores. Because of their very great difference in physiological function, however, it would not be expected that the directional movements of the pigmentary organelles within these bright-colored pigment cells would necessarily be similar to those of the melanosomes of melanophores. Indeed, such a cooperative difference in response is, by nature of their separate functions, essential for maximal color expression of the integument.

Iridophores participate in physiological color change in amphibians (91) and are under hormonal control (91, 134, 135). These cells contract to a spherical contour when frogs are adapted to a black background and, conversely, become stellate when animals are placed on a white background or are hypophysectomized (91). It was early demonstrated that the active principle involved in the darkening response is of pars intermedia origin. Intermedin causes aggregation of reflecting platelets to a perinuclear position within iridophores, and these same organelles redisperse in the absence of this hormone. Iridophore responses are similar in both larval and adult frogs.

Not all iridophores of frogs are responsive to endocrine stimulation. Iridophores, like melanophores, show a physiological response gradient in a dorsal-ventral direction. Both iridophores and melanophores of dorsal frog skin respond much more rapidly and completely to injected MSH or to black-background adaptation than do ventral iridophores and melanophores. Conversely, the effects of MSH wear off much more rapidly on ventral chromatophores than on dorsal chromatophores. Where melanophores are absent and only iridophores are present, as in belly skin, the iridophores do not respond to any endocrine stimulation. These observations imply a gradient of threshold for chromatophore responses and provide a cytological basis for the phenomenon of counter-shading in amphibians, as well as in other vertebrates.

Among amphibians, there is little evidence that xanthophore responses are involved in physiological color change. In most species of amphibians, these chromatophores are always seen to be in the dispersed state and, under these conditions, are partly responsible for the green coloration of frogs. The dispersed state of the pigmentary organelles within xanthophores is unaffected by hypophysectomy or by injections of MSH. Xanthophores in isolated pieces of skin from adult frogs are similarly unresponsive to the presence or absence of MSH. Only in the frog, *Hyla arenicolor*, has the pigmentary component of the xanthophore been observed in the aggregated state. The addition of MSH to isolated pieces of skin from this frog results in pigment dispersion (10).

In one of the last major papers on reptilian chromatophore structure and function, Sand (136) stated that it was not known whether the bright-colored pigment cells "play a dynamic or merely static role in the pigmentary responses of reptiles." In *Anolis,* the xanthophores and iridophores do not appear to respond to hormonal stimulation as they do in some frogs (37). No other observations on the regulation of reptilian bright-colored chromatophores have been reported.

In elasmobranchs, both xanthophores (137, 138) as well as iridophores (139) are present in the integument. According to Hogben (138), both melanophores and xanthophores respond to intermedin in unison. No observations have been made to determine whether iridophores respond to endocrine stimulation.

That xanthophores and erythrophores may be under hypophysial control in teleosts has been known for many years. In fact, the name intermedin takes its origin from the classic papers of Zondek and Krohn (140–143) wherein it was demonstrated that this hormone disperses the pigments within the erythrophores of the European minnow, *Phoxinus phoxinus.* It appears that both xanthophores and erythrophores of teleosts respond to intermedin both *in vivo* and *in vitro.* In *Carassius,* intermedin disperses pigment organelles within both xanthophores and melanophores. It has not yet been established, however, whether intermedin normally plays a role in chromatic regulation in teleosts. It has been demonstrated, however, that prolactin (144) disperses pigment organelles within xanthophores of the fish, *Gillichthys mirabilis.* In this teleost, a pituitary principle is apparently necessary for xanthophore dispersion, for hypophysectomy prevents the normal yellow color change in response to a yellow background. Injections of ovine prolactin into intact or hypophysectomized fish result in a local yellowing of the skin followed by a generalized yellowing of the fish. Both ACTH and MSH are without such a direct effect on the xanthophores. Other evidence strongly suggests that prolactin of this fish's own pituitary plays a role in the normal response to a yellow background (144).

Relatively little is known about the control of iridophore responses among fishes. Most of our information comes from the study of the reflecting cells of the teleost, *Bathygobius* (145). Responses of iridophores to changes in background essentially parallel those of melanophores. On a pale background, iridophores are expanded and melanophores are contracted; the reverse is true on dark-colored backgrounds. Pigmentary responses to background are not markedly altered by hypophysectomy; however, denervation of fin rays leads to the appearance of a dark band in which iridophores are contracted and melanophores are expanded. In such denervated fins, responses of iridophores conform to those of melanophores except that they are in the opposite direction. The role of hormonal factors in controlling physiological responses of iridophores may be subordinate to that of nervous regulation (145).

UNIHUMORAL AND BIHUMORAL
THEORIES OF CHROMATOPHORE CONTROL

For many years there has been great argument as to whether only one hormone, MSH, is involved in vertebrate chromatophore regulation or whether two hormones are involved. The unihumoral theory, as championed by G. H. Parker, favored the view that the control of pigment granule movements was due solely to the presence or absence of circulating intermedin. The bihumoral theory, as elaborated by Sir Lancelot Hogben and his co-workers, involves, in addition to MSH, the presence of a W (whitening; lightening) hormone responsible for causing paling of dark-colored animals by directly overriding the effects of MSH at the effector cell level. Most of the early literature, as well as some of the more recent (146, 147), on pigment cell regulation has been directed toward establishing a firm basis for either the unihumoral or the bihumoral theory of chromatic regulation. Present evidence is completely in accord with a unihumoral theory of vertebrate chromatophore control (148).

The W substance of Hogben was thought to be released from the head region of the amphibian. The two possible sources for the W substance, as suggested by Hogben and Slome (34), were the pituitary gland (pars tuberalis) or, possibly, the pineal gland. The isolation of melatonin, a potent melanosome aggregating agent of dermal melanophores of larval amphibians, by Lerner et al. (149) raised the possibility that melatonin might indeed be the W substance of Hogben and Slome. This is clearly not the case as will become evident from the information presented in the section on the pineal (see Chap. 6). Nevertheless, many students, as well as researchers, have been misled into believing that this pineal agent might well play a normal role in the regulation of vertebrate adaptive color changes (36). It is true that certain hormones can override the effects of intermedin on chromatophores and may even possibly prevent the release of MSH from the pituitary (150, 151). There is no evidence, however, that any hormone other than MSH regulates vertebrate chromatic changes at the effector cell level in response to background adaptation. The ability of poikilotherms to adapt to either a white or a black background can be entirely accounted for by the presence or absence of circulating MSH, or, as in some teleosts, and reptiles by a direct neuronal (sympathetic) control of chromatophores.

REGULATION OF
PARS INTERMEDIA FUNCTION

The question then becomes one of how animals regulate the release of MSH in adaptive color changes. The control system involved, as for other hormones of the pituitary, has been shown to reside in the hypothalamus. Etkin (152) provided evidence that the intermediate lobe of the frog or tadpole adapted to a light-colored background is under an inhibitory influence originating in the infundibular stalk (hypothalamo-hypophysial nerve tract). The hypothesis was advanced (153) that the activity of the cells of the intermediate lobe may, like those of the distal lobe, be controlled through a neurohumoral mechanism. Lesions or other damage of the hypothalamus causes darkening in a number of species of amphibians (152, 154, 155), suggesting that the neurosecretory fibers of the hypothalamo-hypophysial fiber tract might be responsible for the normal inhibition of the pars intermedia. It was proposed that the hypothalamus controls the secretion of intermedin by way of inhibitory neurosecretory nerve fibers entering directly into the substance of the gland (155–157).

Electrical stimulation of the hypothalamus of the frog (*Rana pipiens*) results in a release of intermedin from the pituitary (158), indicating that the neural elements in the hypothalamus which normally inhibit MSH secretion are themselves inhibited by such stimulation. The pattern of events during the reestablishment of control of pars intermedia function after sectioning of the nerve supply in amphibians suggests that this regulation is mediated by ordinary nerves, not neurosecretory nerves (159). Furthermore, injections of minute amounts of certain pharmacological agents into the frog's brain indicate that the hypothalamus is the area of the brain apparently involved in the regulation of MSH release (160). Both a cholinergic and an adrenergic nervous pathway inhibitory to MSH release have been demonstrated. Since acetylcholine is very effective in releasing MSH, it was suggested that an inhibitory cholinergic pathway is linked directly with a second pathway (adrenergic), which is also inhibitory. The release of acetylcholine from cholinergic axons was suggested to regulate (inhibit) adrenergic neurons whose release of norepinephrine (a catecholamine) would be inhibitory to MSH release. The adrenergic component of hypothalamic inhibition was thus considered functionally closer to the pars intermedia (Fig. 5–6). A cholinergic mechanism of MSH regulation in urodele larvae has also been suggested (161).

By means of a sensitive fluorescence technique, a plexus of aminergic (catecholamine-containing) nerves in the pars intermedia of the frog, *Rana temporaria*, has been demonstrated (162). This plexus, which "synaptically enclosed" the intermedia cells, was said to possess the basic char-

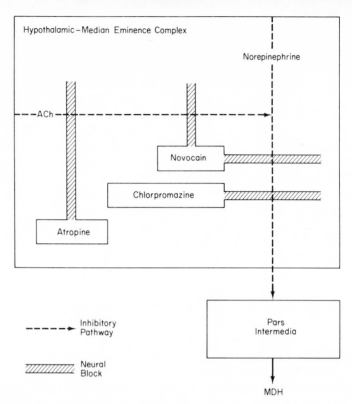

FIG. 5-6. Hypothetical scheme for the hypothalamic regulation of pars intermedia function in the frog. [Redrawn from (160)]

FIG. 5-7. Sagittal section showing part of the pars intermedia and adjacent lobes of the pituitary of the adult mouse. **A.** Neural lobe with only a few fluorescent particles near pars intermedia. **B.** Pars intermedia; note the weak fluorescence of the cytoplasm of the cells and the intensely fluorescent varicose fibers. **C.** Pars distalis. [From (164)]

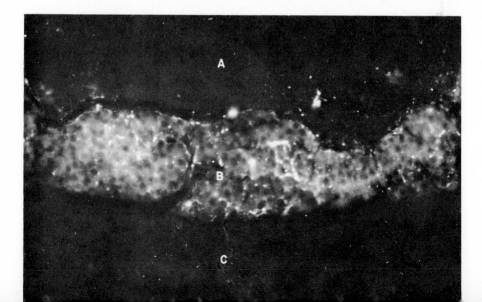

acteristics of an adrenergic ground-plexus. A similar plexus has also been found in the pars intermedia of the rat (163), the mouse (164) (Fig. 5–7), and other mammals. These observations provide strong support for the view that the fluorescent nerves in the pars intermedia of the frog are adrenergic synaptic terminals, a catecholamine being the neurohumoral transmitter agent. The fact that reserpine blocks the inhibition of the pars intermedia (and depletes the fluorescent plexus in the pars intermedia) supports the hypothesis that adrenergic neurons may mediate the inhibitory control, rather than, or in addition to, neurosecretory neurons. The origin of these inhibitory neurons in the brain is unknown.

Nerve endings containing both clear and dense "synaptic vesicles" in close contact with the colloidlike vesicles of the parenchymal cells are abundant, as seen by the electron microscope in the pars intermedia of the toad, *Bufo arenarum,* (165). The dense synaptic vesicles were interpreted as vesicles containing norepinephrine. On this basis it was speculated (166) that the control by the hypothalamus of the pars intermedia of amphibians might be regulated through non-neurosecretory fibers of adrenergic nature. Similar observations of the pituitary of a teleost, the Argentinian eel, *Conger orbignyanus,* suggests that an inhibitory hypothalamic control of the pars intermedia may be, as in amphibians, exerted by adrenergic neurons (167).

Transplantation of the neurointermediate lobe of the pituitary of the toad, *Bufo bufo,* into the anterior chamber of the eye (where high concentrations of catecholamines are present) results in what appears to be a submaximal secretion of MSH (168). Intermedin secretion becomes maximal after extirpation of the ipsilateral first cervical sympathetic ganglion innervating the eye. If the eye containing the graft is removed, the integumental melanophores show a maximal contraction. When dopamine or norepinephrine is injected into the anterior chamber of an eye bearing a pars intermedia graft, there is a drop in the melanophore index that is not duplicated by similar injections into the contralateral eye (169). This fact was interpreted as additional evidence that adrenergic nerve fibers exert an inhibitory control on the secretion of MSH.

Transplantation of the pituitary to an ectopic site usually results in a somewhat diminished functioning of the vertebrate pars distalis. There is, in contrast, a hypersecretory activity of the pars intermedia after such transplantation, as observed microscopically (170) at the pars intermedia level or cytologically at the effector cell level (156). Hypothalamic lesions in adult frogs (*Rana nigromaculata,* 171) or the removal of the anterior hypothalamus in tadpoles (*Xenopus laevis,* 155) results in hypertrophy of the intermediate lobe and a decrease in pituitary MSH (171, 172), which is accompanied by hyperpigmentation of the integument (Fig. 5–8) (173). Although direct measurements of circulating levels of MSH in frogs have

A

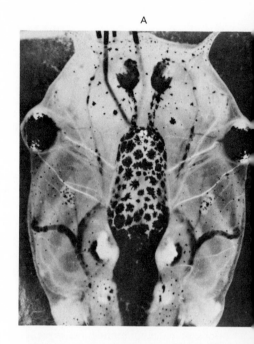

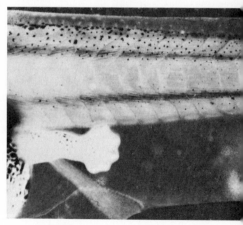

FIG. 5-8. Head (**A–C**) and tail (**D–F**) regions of tadpoles of the African Clawed Toad, *Xenopus laevis*. Compare the state and number of melanophores in the normal control (**B** and **E**) to that of a tadpole raised on a white background (**A** and **D**) or to that of a tadpole in which the hypothalamic inhibition of MSH release has been removed by extirpation of the prosencephalon (**C** and **F**). [From (173)]

D

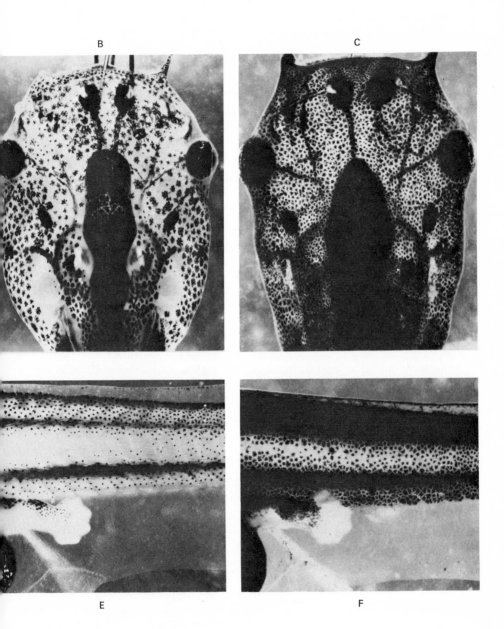

not been made, hypophysectomized frogs parabiosed to hypothalamic-lesioned frogs become dark in color, thus indicating the presence of a circulating chromatophorotropin (172).

In a number of electron microscopic studies (174–176), specific cytological changes in pars intermedia cells could be correlated with the nature of the background to which frogs are adapted. Pars intermedia cells of frogs adapted to a black background contain an extensive endoplasmic reticulum but only a few secretory granules, which might be taken as indicative of an active hormone synthesis and release (176). In contrast, pars intermedia cells of frogs adapted to a white background contain a less-extensive endoplasmic reticulum but numerous secretory granules, indicating synthesis but little or no release. These observations correlate with bioassays of the frog pars intermedia after background adaptation. There is more MSH present in the pars intermedia of frogs (*Xenopus laevis,* 177) adapted to a white background than in the pars intermedia of frogs adapted to a black background. Other workers, however, have failed to find such a difference in pars intermedia content of MSH in the frog, although cytological differences have been noted (178).

All these results clearly establish that MSH secretion from the pars intermedia is under inhibitory regulation, apparently by the hypothalamus. Most interesting in this regard, therefore, is the observation (179) that weasels, *Mustela erminea,* with autographs of the pituitary under the kidney capsule grow only brown hair, even when exposed to stimuli that induce growth of the white coat in intact animals. Thus in the mammal, as in the amphibian, the hypothalamus exerts an inhibitory influence over the secretion of MSH. Since hypothalamic lesions in the rat may result in either a hyperactive or a hypoactive pars intermedia (180), it has been suggested that a dual inhibitory and excitatory system may control pars intermedia activity in the rat (180, 181). A similar dual regulation of pars intermedia control in other vertebrates had earlier been suggested.

In the selachian (*Scylliorhinus stellaris,* 182), electron microscopy revealed that two types of neuronal endings are present within the pars intermedia. On the basis of the structural characteristics of the contained granules within the neural endings, it was suggested that one type of neuron is neurosecretory in nature, whereas the other is adrenergic. Both types of neurons synapse with individual pars intermedia cells; and it was suggested that the neurosecretory neuron regulates MSH synthesis, whereas the adrenergic neuron regulates MSH release. Both types of neurons were considered to be stimulatory in nature. Other ultrastructural observations on the skate, *Raja radiata,* pars intermedia confirm the presence of both neurosecretory and adrenergic neuronal elements in close contact with pars intermedia cells, but the possible individual roles of each neuronal type appear more complicated than previously suggested (183). Nakai and Gorbman (184) similarly demonstrated that at least two different types of nerve synapses may end upon the individual parenchymal cells of the frog pars intermedia. Again, this supports a hypothesis of a doubly innervated secretory unit.

Although ultrastructural studies have consistently demonstrated both the presence of adrenergic neurons and their synaptic contact with pars intermedia cells of the frog, the so-called neurosecretory neurons are sparse in number and restricted to the border of the pars intermedia adjacent to the pars nervosa. Whether these large-granule-containing cells are really neurosecretory in nature and play an active role in pars intermedia function is in need of further study. Since neurosecretory-like axons are limited to the frog pars intermedia adjacent to the pars nervosa, it is difficult to accord them a role in pars intermedia regulation other than by a mechanism involving a release of a diffusible regulatory substance (184). Numerous nerve end-

FIG. 5-9. Pituitary of the lizard, *Klauberina riversiana.* Unlike the pars intermedia (PI) of other nonreptilian vertebrates, the pars intermedia of this lizard is completely negative to aldehydefuchsin, a stain generally used to stain for neurosecretions. The arrow at right points to fuchsinophilia, which is within a process of the vascular septum that separates the intermediate lobe from the darkly stained neural lobe (NL). The hypophysial cleft (HC) separates the pars intermedia from the pars distalis (PD). [From (186)]

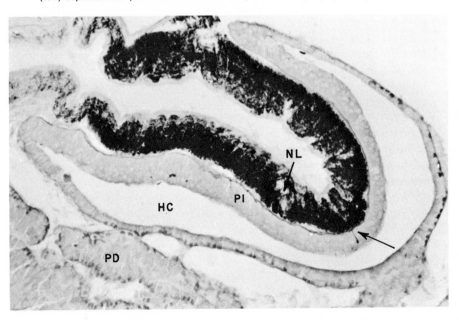

ings making synaptic contact with the pars intermedia cells of the mammalian pituitary have been noted (185). Here, too, adrenergic axonal contacts are more prominent than are neurosecretorylike axonal synapses. It is interesting in this context, therefore, to note that initial observations of the ultrastructure of a reptilian (the lizard *Klauberina riversiana*) pars intermedia (Fig. 5–9) reveal the total absence of axonal endings (186). It is possible, as suggested for the regulation of other pituitary hormones (187), that neurosecretory substances from the median eminence or from pars nervosa axonal endings might regulate pars intermedia function (186).

Electrophysiological studies (188, 189) indicate that two types of spontaneously active electrical units are present in the pars intermedia of the frog. These units are neuronal in nature; one neuron is inhibited by light, whereas the other is indifferent to changes in illumination. The light-indifferent neuron was considered tonically inhibitory and, in harmony with past experimental data, adrenergic in nature. The light-inhibitable neuron was considered to stimulate MSH release under conditions of low illumination, when its electrical activity is maximal, by inhibiting the tonically inhibitory neuron. Under conditions of strong illumination, when its electrical activity is minimal, the light-inhibitable neuron was considered inactive with respect to inhibiting the tonic influence of the inhibitory neuron. Under such conditions, MSH would not be released.

According to this scheme, the two nervous elements in the pars intermedia are in balance and regulated by the influence of light on the activity of one of them. Whether the light-inhibitable neuron corresponds to the neurosecretory or cholinergic neurons that have been implicated in pars intermedia control is unknown. Since axo-axonal synapses were found in the pars intermedia of the frog (184), it is possible that regulation of pars intermedia function could be mediated through inhibition of one neuron by another, in addition to any dual regulatory role that might be involved directly at the effector cell level. The proposed cholinergic link in the regulation of pars intermedia cell function might relate to the fact that this neurotransmitter agent is apparently present within neurosecretory axonal endings and may be necessary for the release of the large granules from the neurosecretory axons (190).

In both the mammal and the frog, it has been shown that the hypothalamus contains both MSH–inhibiting (191–195) and MSH–releasing factors (194, 196). The possible relationship of these hypothalamic factors to pars intermedia regulation as previously discussed is unclear. Since an adrenergic mechanism of pars intermedia control has only been suggested for poikilotherms, it is possible that these hypothalamic regulatory factors may be of particular importance to the mammal, in which they have been most extensively studied. It may be that an adrenergic mechanism of pars intermedia control would be important in some poikilothermic vertebrates in which rather rapid chromatic adjustments to background are important. On the other hand, hypothalamic factors similar to those that regulate the release of other hypophysial hormones in mammals may play a dominant or additional role in all vertebrates wherein slower seasonal (morphological) color changes are involved. At present it is impossible to evaluate the relative roles of adrenergic, cholinergic, and neurosecretory neurons in the regulation of pars intermedia function.

chromatophore control II

general endocrine
and nervous mechanisms
of chromatophore control

PINEAL ROLE IN THE CONTROL
OF VERTEBRATE
CHROMATOPHORE RESPONSES

As early as 1911 von Frisch (197) observed that the pineal or closely related diencephalic area of the teleost, *Phoxinus laevis* (= *Phoxinus phoxinus*), was sensitive to light and influential in controlling melanophore responses. Babak (198) and Laurens (199) showed that larvae of several amphibian species are paler when placed in the dark than when they are under conditions of illumination. Fuchs (200) felt that larvae produce, as a result of their own life processes, certain substances that cause melanosomes to aggregate. Thus, from internal stimulation, their melanophores contract under the influence of these endogenous metabolic substances. When light is provided, however, the pineal is stimulated, which results in an inhibition of the endogenously induced melanosome aggregation. Laurens (201) opposed this view and, in an extensive study involving localized illumination of the pineal area of the head of larval urodeles (*Ambystoma opacum* and *Ambystoma maculatum*), concluded that the epiphysis has no influence on the reactions of larvae to light and to darkness. His arguments were so convincing that McCord and Allen (202), who discovered that extracts of beef pineals cause blanching in *Rana pipiens* tadpoles, concluded that "while the pineal does not act in the role of its ancient ocular function, it contains within itself an active principle capable of inducing pigment changes independent of and wholly apart from environmental conditions."

Scharrer (203) confirmed the earlier observations of von Frisch, and Young (123) observed that the pallor which occurs in larval lampreys on transference from light to darkness is abolished after removal of the pineal complex. Bors and Ralston (204) observed that pineal extracts of both pig and man induce melanosome aggregation in larval and adult *Xenopus laevis*. Lerner et al. (149) isolated a potent melanosome–aggregating agent from beef pineal glands. They identified this compound as melatonin (*N*-acetyl-5-methoxytryptamine). See Fig. 6–1.

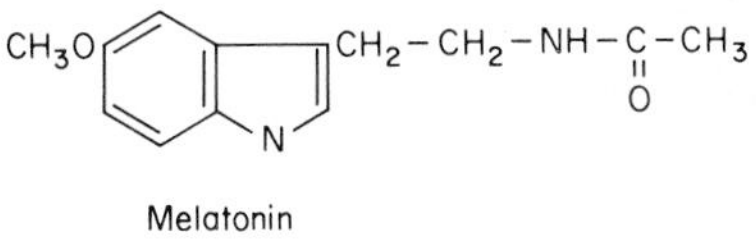

FIG. 6-1. Structure of melatonin.   Melatonin

As a result of the discovery that the tail-darkening reaction of *Xenopus* larvae (86, 205), an event that occurs when larvae are placed in the dark (Fig. 6–2), is mediated by the direct action of light on tail-fin melanophores, it was considered possible that a similar photochemical mechanism might mediate the body-lightening reaction. The temporal events in body lightening and subsequent recovery do not, however, seem consistent with a direct control of melanophores by a photochemical system (Fig. 6–3). Instead, they imply that a hormonal mechanism is involved, with relatively rapid onset of melanosome aggregation corresponding to release of a stored hormone and slow recovery concordant with gradual loss of this principle from the circulation. With the implication of a hormonal mechanism in the body-blanching reaction, suspicion arose that perhaps the melanosome-aggregating principle of the pineal is the active agent. This suspicion grew even stronger because of the fact that the paling reaction occurs when eyeless larvae are placed in darkness (199, 206). It seemed possible, therefore, that the pineal was the photoreceptor necessary for the paling reaction. Following this line of reasoning, an hypothesis was established that explained the mechanism of the body-blanching reaction completely in terms of two aspects of pineal physiology—photoreception and endocrine function. Briefly, the hypothesis states that when amphibian larvae are placed in darkness, the pineal is stimulated by the absence of light, thus causing it to produce and release a melanosome-aggregating agent and resulting, finally, in the blanching reaction. The first publication of this hypothesis (206) included data showing that pinealectomized *Xenopus* larvae are not capable of performing the blanching reaction when placed in the dark.

Implicit in the hypothesis that the pineal is responsible for body blanching is the presence of photoreceptors in some part of the pineal complex. In the amphibian pineal, photoreceptive structures have been found that resemble those found in retinas of vertebrate lateral eyes (207). In addition, the frontal organ (Stirnorgan, "brow spot"; 208) and its reptilian

FIG. 6-2. *In vitro* tail-darkening response of isolated tails from *Xenopus laevis* larvae under normal conditions of illumination (top) and in darkness (bottom). Only the melanophores in the tip of the tail (center) respond to darkness if the rest of tail is illuminated. [From (86)]

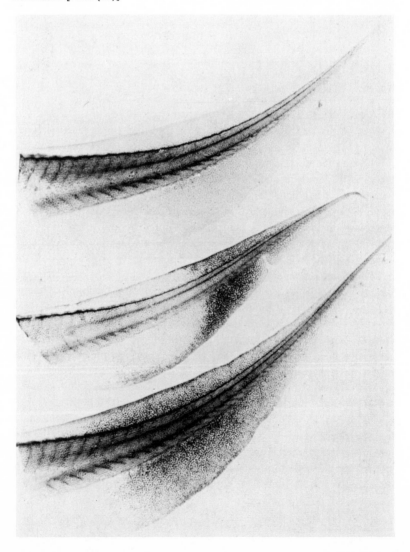

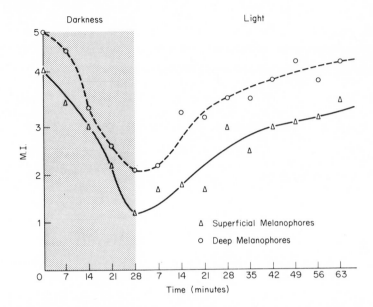

FIG. 6-3. Melanophore responses, expressed as a melanophore index (MI) (34), of *Xenopus laevis* larvae (stages 48–49) during development of, and recovery from, the body-lightening response induced by a 28-minute exposure to darkness (shaded area). [Redrawn from (219)]

homologue, the parietal eye (209–211), both possess cells that are structurally similar to those of photoreceptors of the retina of the lateral eyes. Both organs are connected to the brain by a nerve or pineal tract (212) that passes dorsally through the epiphysis (Fig. 6–4). The photoreceptive elements within these organs appear to be functionally photoreceptive in that they are capable of wavelength discrimination as determined by measurements of the electrical activity of the surface of the parietal eye itself (213) or from the efferent nerves from the Stirnorgan (214). Whether these organs regulate either pineal or other brain function by means of direct efferent nervous pathways or by secretions is not known. Removal of the Stirnorgan in *Xenopus* has no effect on the ability of either larval or adult frogs to background adapt (215–217) or to blanch in the dark (218). The mammalian pineal lacks photoreceptive elements, but it is regulated by a nervous pathway from the retinal photoreceptors of the lateral eyes.

A salient feature of the photoreceptive aspect of the blanching reaction is that the response occurs as a result of lack of illumination. Undoubtedly this is one reason why Laurens (201) discounted any influence of light on the pineal organ. He attempted to obtain an active melanophore response by illuminating the pineal directly, thinking that this structure

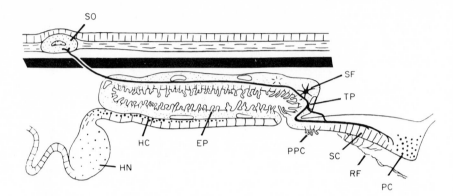

FIG. 6-4. Diagrammatic sagittal section of the organs of the anuran diencephalic roof. SO, Stirnorgan; HN, habenular nucleus; HC, habenular commissure; EP, epiphysis; SF, secretory follicle; TP, tractus pinealis; PPC, posterior pineal cluster; SC, subcommissural organ; RF, fiber of Reissner; PC, posterior commissure. [From (212)]

should react as a result of a positive photostimulus. The concept that the pineal is affected by a lack of light stimulation has gained support from other experiments (214, 218). After removal of the lateral eyes and the Stirnorgan, photic responses were recorded from a very localized region in the diencephalic roof of the frog's brain. Systematic exploration with a recording electrode indicated that the photic responses originated from the epiphysial stalk. A sustained discharge of action potentials was recorded from the epiphysial stalk in the dark. This activity was inhibited by light of all wavelengths.

Evidence that the pineal releases a hormonal agent which is responsible for the blanching reaction comes from several sources. The first direct experiments (206, 219) involved cautery of the diencephalic roof. Older *Xenopus* larvae "pinealectomized" in this manner consistently fail to blanch when placed in the dark. Similar results were obtained by others working with salamanders (*Ambystoma opacum,* 207, 220; and *Taricha torosa,* 207).

All the experiments described so far refer to mediation of the blanching reaction by a "pineal hormone." The identity of the specific hormone active in this response is unknown; however, it has been suggested (206, 219) that it is melatonin. At the moment, no other hormone appears so likely a candidate as melatonin, the physiological effects of which fit all the prerequisites for the blanching reaction.

Of direct importance is the fact that melatonin is a powerful melanosome-aggregating agent. The minimal effective dose of melatonin required for aggregation of larval *Xenopus laevis* or *Rana pipiens* melanosomes is 0.0001 μg per milliliter of water in which the tadpoles swim (36, 219). No other agent so far tested is nearly as effective (221). On the basis of several observations, including the occurrence of body blanching in the face of hypophysial chromatophore stimulation, the rapid onset of the blanching reaction, the relatively short period required for inactivation of the "pineal hormone" during recovery from the blanching reaction, and the occurrence of the tail-darkening reaction concomitant with blanching, it seems that the natural melanosome-aggregating agent is one that is active at a very low concentration. Melatonin certainly fulfils this requirement. It should be noted also that the character of pallor induced by melatonin is identical to that which occurs during normal blanching. Deep melanophores on blood vessels, nerves, and various organs, as well as those in the integument, contract markedly. Moreover, the temporal response of *Xenopus* melanophores to melatonin bears striking resemblance to the naturally occurring blanching reaction (222).

Experiments utilizing hypophysectomized tadpoles of both *Rana pipiens* and *Hyla arenicolor* show clearly that the control of the blanching reaction does not operate through the hypophysis. As a result of the lack of a pituitary chromatophorotropin in such larvae, melanophores are contracted and iridophores are expanded. When hypophysectomized tadpoles are placed in the dark, melanosomes that are already aggregated become even more densely aggregated within melanophores. A clearer picture of this reaction is obtained by immersing hypophysectomized tadpoles in water containing MSH. This action results in a prominent melanosome dispersion. When these MSH-treated, hypophysectomized tadpoles are placed in the dark, while others are left under normal illumination as controls, those tadpoles placed in the dark blanch, whereas their controls retain melanophores with dispersed melanosomes. Clearly, then, the mechanism of the blanching reaction does not seem to involve an inhibition of the release of a chromatophorotropic hormone from the hypophysis, for the reaction occurs perfectly well in the absence of the hypophysis. These experiments also seem to reveal that in the blanching reaction of normal larvae, the "pineal hormone" overrides the melanosome-dispersing stimulation of the endogenous chromatophorotropic hormone *at the effector cell level.*

Additional evidence to support such an interpretation comes from the observation that epidermal melanophores remain dendritic during the body-lightening response of *Rana pipiens* tadpoles (223, 224). These epidermal melanophores do have the capacity to aggregate their melanosomes, and they do so in the absence of intermedin but not in response to melatonin. Thus this observation is good evidence that intermedin is present during the body-blanching response. Additional support is also available from studies on both larval and adult amphibians. Melatonin has no effect on iridophores when injected into frogs adapted to a black background or on iridophores of frog skins darkened by MSH, *in vitro*. The very minimal lightening effects of melatonin on frog skins, either *in vivo* or *in vitro* are due to the slight melanosome aggregation occurring within dermal melanophores. Here the maintenance of the aggregated state of the iridophores indicates that the effects of intermedin still persist. Since neither iridophores nor epidermal melanophores of either adult (36) or larval amphibians (151, 224) are responsive to melatonin, *in vivo* or *in vitro*, lightening effects of this agent on adult frogs *in vivo* must be ascribed to its direct effects on dermal melanophores rather than to any effects on MSH release, as suggested by Charlton (217).

It is of historical interest to note that McCord and Allen (202) clearly demonstrated that the lightening effects of pineal extracts on tadpole melanophores were on dermal melanophores and not on epidermal melanophores. Both Smith (85) and Atwell (102) confirmed the observations of McCord and Allen (202), and, in addition, they noted that although immersion of tadpoles in pineal extracts caused contraction of dermal melanophores, the iridophores were unaffected and remained contracted. These early observations are consistent with our present observations today using pure melatonin (21, 225).

All these observations indicate that although there is strong evidence that melatonin may play a natural role in the control of chromatic changes (dermal melanophore responses during the body-blanching response) in young larval amphibians (206), no such role has been demonstrated for adult amphibians (250). Bogenschütz (251) found that epiphysectomized adult *Rana esculenta* still become pale when transferred from a black to a white background. Histological examination of the midbrain confirmed that the pineal was entirely destroyed and no regeneration had taken place. These results would appear to further rule out a role for the pineal in chromatic background adaptations in older-stage anurans.

Oshima and Gorbman (189) found that the integrity of the pineal organ is required for light inhibition of the light-sensitive neuron of the frog pars intermedia, suggesting that this organ may be the initial light receptor for pars intermedia regulation. The eye was found not to be the photoreceptor. This interesting observation, although inconsistent with other theories on the role of the pineal and the lateral eyes in chromatic regulation in amphibians and other poikilotherms, is important, for a role has been proposed for the pineal in the regulation of coat color in a mammal (226).

Pinealectomy increases the MSH content of the rat pituitary, whereas injected melatonin decreases it (227). Since constant illumination, which suppresses melatonin formation, increases pituitary MSH levels, it was considered "logical to assume that the pineal acting via melatonin or other of its secretions, mediates the effect of light upon MSH release" (228). Since the MSH-inhibiting factor (MIF) of the mammalian hypothalamus increases pituitary MSH content (229), the suppression of melatonin release by constant illumination was considered (227) to have "eliminated the mechanism that opposes the action of MIF." It should be noted that there is no known effect of melatonin on pigmentation in the rat.

Melatonin implanted into male weasels, *Mustela erminea bangsi,* results in brown or white weasels undergoing the spring molt to the brown pelage to grow a new white coat, whereas control animals retain or acquire the brown coat (226). Melatonin-treated as well as nontreated (control) weasels with pituitary autografts under the kidney capsule grow brown hair after hair growth is initiated by plucking. This fact suggests that melatonin does not act directly on the pituitary gland to inhibit MSH secretion. Rather, it was postulated that melatonin acts on the hypothalamus, causing the release of the MSH-inhibiting factor, which, in turn, prevents MSH release. Obviously the proposed roles of melatonin in the rat and the weasel are contradictory and therefore provide no easy interpretation for a possible pineal role in mammalian pigmentation.

These are, nevertheless, important observations, and they suggest a possible role for the pineal and melatonin in coat color regulation in mammals, particularly those undergoing seasonal pelage color changes. Additional confirmatory data, however, are needed for other mammals. Since melatonin implantations result in reproductive quiescence and have an inhibitory effect on initiation of hair growth in the weasel, it is possible that the effects of melatonin may be indirect. Other investigators have failed to effect pigmentary changes in other mammals (230), including man (231), with massive injections of melatonin, and mouse melanophores are refractory to melatonin *in vitro* (232).

Whether the pineal gland and melatonin are similarly involved in the control of pigment cell responses in other vertebrates is uncertain. McCord and Allen (202) described experiments by Noble which showed that the melanophores of larval *Fundulus heteroclitus* become contracted following addition of pineal extract to the medium, whereas those of adults are unresponsive. Wyman (233) observed melanosome aggregation within melanophores of embryonic and larval *Fundulus* but failed to note any effect of pineal extracts on the melanophores of adults (234). Hewer (235) was also unable to cause melanophore contraction in adult *Phoxinus* injected with pineal extracts.

Melatonin is an effective melanosome-aggregating agent of embryonic melanophores of *Fundulus* and, to a limited extent, the melanophores of larvae (236). The melanophores of adult *Fundulus,* in contrast, are totally insensitive, *in vitro* (237) or *in vivo* (236), to melatonin. These results using melatonin are thus similar to those obtained with pineal extracts by earlier investigators.

What is of interest in these experiments is the apparent developmental change in response of melanophores of *Fundulus* to melatonin. Somewhat similar results were reported for *Rana pipiens* (151). The dermal melano-phores of *Rana pipiens*, like those of *Xenopus* larvae, are exceedingly sensitive to melatonin, but there is a decline in this sensitivity with developmental age. Similarly, McCord and Allen (202) found that after metamorphosis the melanophores of *Rana pipiens* were no longer responsive to pineal extracts. Developmental changes in vertebrate melanophore control appear to be reflected in these differential responses of embryonic, larval, and adult melanophores to melatonin.

There is presently little or no evidence that melatonin plays any role in the normal control of chromatic responses in teleosts. In those few studies where melatonin has been reported to have an effect on chromatic responses in teleosts (238, 239), it is yet to be established whether such effects are of physiological rather than pharmacological importance. There are but few data in support of a view that the teleost pineal gland releases a hormonal agent, such as melatonin, that might control teleost melanophore responses. Melanophores of adult *Fundulus* become contracted when fish are placed in the dark (240, 241). The same response will take place in hypophysecto-mized individuals (241), thus indicating that the response is not under the control of the pituitary gland. Blinding, by removal of the eyes, results in a darkening of the skin of most teleosts, including *Fundulus*. The darkening —for example, in *Fundulus* (234) and the catfish, *Ameiurus nebulosus* (242) —is followed by a contraction of the expanded melanophores when these blinded fish are then placed in the dark, thereby indicating that the eyes are not involved in the latter response. Denervated melanophores of blinded fish (*Ameiurus nebulosus*; no data on *Fundulus*) remain expanded in the dark, at least initially, in contrast to the contracted state of the innervated melanophores (242). Pinealectomized *Fundulus* still exhibit a blanching response to darkness (243). Pinealectomy combined with blinding in the trout, *Salmo gairdneri*, does, however, modify slightly the nocturnal blanch-ing (244). Melanophore contraction in response to darkness was considered in *Phoxinus* (197) and *Fundulus* (234) to be under the control of a photo-receptor in the brain. Both von Frisch (197) and Scharrer (203) suggested that the light receptor (in *Phoxinus*) was situated in the vicinity of the diencephalon, but they were unable to locate it precisely; light sensitivity was not completely abolished by pinealectomy. Electron microscopic studies (245) on the pineal organ of *Phoxinus* demonstrate the presence of sensory cells strongly resembling photoreceptors. These observations provide a good anatomical basis for the light sensitivity of the diencephalon of blinded minnows as described by von Frisch (197) and Scharrer (203).

It would appear that light stimulation of the pineal and/or associated areas of the brain may have contrasting effects, depending on the species involved. Von Frisch and Scharrer demonstrated that light stimulation of the pineal area of *Phoxinus* resulted in melanosome dispersion within melanophores. In contrast to this observation, it has been noted (246) that melanosomes become dispersed within melanophores when the pineal region, of a number of species of teleosts, is covered with India ink, thus apparently preventing light stimulation of this area. That photostimulation of the pineal area can evoke two diametrically opposed types of melanophore response argues strongly against any suggestion that light stimulation of the pineal area leading to pigmentary responses is mediated by release of a humoral agent (such as melatonin) of pineal or closely associated diencephalic origin. The fact that the blanching response to darkness and the reverse response to light stimulation are rapid events that can be repeated in succession over a long period of time is also strong evidence against a humoral agent being the factor in melanophore control of the blanching response.

If the pineal and/or adjacent areas of the brain of teleosts are involved in the blanching reaction that takes place in the dark, then the response is controlled, as suggested by von Frisch (197), through nervous pathways that eventually lead directly from such areas to the melanophores themselves. The response cannot be due to a release of a circulating hormone, such as melatonin, as may take place in larval amphibians. Nor can the contraction of melanophores under normal lighting conditions (when fish are placed on a white background) or in the dark be attributed to a pineal-pituitary axis involving the inhibition of intermedin release from the hypophysis, for the responses take place in hypophysectomized fish (241). The failure of denervated melanophores to contract rapidly in the dark, as do other innervated melanophores, clearly indicates that the absence of light stimulation of the central nervous system results in a lightening of fish that is neuronally controlled rather than humorally regulated. Decapitation of *Fundulus* results in an immediate expansion of all melanophores, thus clearly implicating a nervous rather than humoral control of melanophores (234).

Although a prominent parietal eye may be present in some lizards, and a pineal organ is present in most reptiles, except crocodiles, there is no evidence that these organs play a role in chromatic regulation in reptiles. Like most amphibians and fish that have been observed, the few reptiles that have been studied blanch in the dark. Under conditions of strong illumination, the lizard, *Anolis carolinensis,* is a dark brown color on a black background but becomes a light (green) color on a white background or when transferred to the dark. Blinded lizards are dark in color under conditions of strong illumination but become pale in the dark; thus the eyes do not regulate this response. Parker (247) noted a similar response with *Phrynosoma blainvilli,* as have numerous other investigators working with chameleons (136); and Parker demonstrated that the response took place whether the parietal eye is covered or not. Carlton (248) noted that *Anolis* is brown if the head is placed within a black box but the body is still subjected to illumination. The pituitary is necessary in *Anolis* for the color change when lizards are transferred from darkness to conditions of illumination (117). The time for paling in darkness suggests that melanosomes aggregate when circulating pituitary chromatophorotropin is absent.

In *Anolis,* melatonin has only a minimal and inconsistent lightening effect *in vitro* on MSH-darkened skins and is ineffective *in vivo.* Thus no experimental data exist at the moment to support a view that the pineal or a pineal hormone, possibly melatonin, plays any role in chromatic regulation in *Anolis.* The suggestion by Kleinholz (117) that integumental photoreceptors reflexly regulate the release of a hypophysial chromatophorotropin responsible for controlling melanophore responses is the only present explanation for the chromatic responses of blinded lizards. The physiological basis for the paling response of blinded lizards (*Anolis*) in darkness is in need of further study.

In elasmobranchs, blinded pups (*Mustelus canis*) are slightly lighter in the dark than in the light, and this condition is due to a melanophore response (1). Melanophores are maximally contracted in hypophysectomized individuals whether in the light or in the dark, thus indicating that melanophores do not expand in response to light. Whether the pineal influences the paling response in the dark has not been studied.

In both larvae and adults of the cyclostome, *Lampetra planeri,* there is a regular daily rhythm of color change (123). Animals are dark in color during the day and become pale during the night. Continuous illumination stops the rhythm and the larvae remain dark. In continuous darkness, the diurnal changes are greatly diminished or the animals become permanently dark. Hypophysectomized larvae become maximally pale regardless of the nature of the illumination but do darken in response to injected pituitary extracts. Removal of the pineal complex (pineal and parapineal bodies) from larvae stops the diurnal rhythm and the animals remain dark under all conditions of illumination. Removal of the pineal from adults only partially disturbs the diurnal color changes, but the rhythm is totally abolished if the paired eyes are removed.

These results were interpreted to indicate that paling of larvae when passing from conditions of illumination to darkness is probably due to an inhibition of pituitary secretion by nervous impulses originating from the pineal complex. This explanation was chosen, rather than an alternate one involving the liberation of a melanophore-contracting or a pituitary-inhibiting substance. The former hypothesis was chosen over the latter mainly because pinealectomy of hypophysectomized larvae did not result in a darkening of the animals, which was expected to happen, according to Young, if the pineal produces a melanosome-aggregating substance. No interpretation of the role of the lateral eyes in adult *Lampetra* was provided. There is no direct evidence to support or refute the suggestion that a direct neural pathway from pineal complex to pituitary regulates MSH release in the ammocoetes larvae. Young's data (123), however, could just as easily support the hypothesis that a humoral agent, such as melatonin, is released from the pineal or associated epithalamic structure and is responsible for either inhibiting MSH release or directly antagonizing MSH at the effector cell level.

The melanophores of the larvae of the lamprey, *Geotria australis,* are exceedingly sensitive to injected melatonin, and implantations of a pineal complex from *Geotria* induce a localized pallor, thereby demonstrating the direct response of melanophores to such stimulation (124). *Geotria* larvae undergo a diurnal rhythm of color change, and this process is abolished by pinealectomy. The melanophores of another lamprey, *Mordacia mordax,* in contrast, are insensitive to melatonin, and the pineal complex is poorly developed; its removal has no effect on chromatic responses.

In birds, there is some indirect evidence that melatonin may be present in the pineal (249). Neither the pineal nor melatonin, however, has been implicated in the control of color change in this vertebrate class.

Altogether these experimental observations reveal that both larval and adult poikilothermic vertebrates lighten on transference from the light to the dark. Not unexpectedly, however, a number of quite different mechanisms for doing so seem to have evolved. The pineal and/or associated area of the epithalamus has been especially implicated in the regulation of this body-blanching response in more than one vertebrate group.

DIRECT NEURONAL
CONTROL OF CHROMATOPHORES

As early as 1852 Brücke (252) provided data suggesting that color change in the chameleon was controlled by nerves. Pouchet, in 1876, demonstrated (253) that sectioning of the peripheral nerves of a teleost led to darkening in the denervated areas and that electrical stimulation of spinal nerves to the skin induced pallor. There is strong evidence that teleost melanophores are controlled, at least in part, by the autonomic nervous system.

Ballowitz (254) described a network of nerve fibers associated with teleost dermal melanophores. Spaeth and Barbour (255) showed that epinephrine caused a rapid and maximal contraction of the teleost melanophore. Numerous other investigations have clearly shown that catecholamines, such as epinephrine (adrenaline) and norepinephrine (noradrenaline), are potent contracting agents of teleost melanophores. Present evidence favors the view that melanophore contraction is controlled through adrenergic neurons that either directly mediate their effects or act on a catecholamine store (256) near the melanophore effector sites. This view is strongly supported by the observation that tyramine, which is known to release norepinephrine from nerve endings, causes melanophore contraction but not after reserpine treatment, which is known to deplete neuronal catecholamine stores (256). The dermal melanophores of the rainbow trout (*Salmo gairdneri*) are intimately enclosed by a dense network of nerve fibers that by fluorescent histochemical criteria were considered catecholamine-containing in nature (257). These nerves could take up and concentrate norepinephrine. If the nerves were severed, the melanophores expanded and there was a disappearance of the fluorescence from the nerves. Similar observations of histological fluorescence have been made on another teleost (*Tautogolabrus adspersus,* 258). According to Falck et al. (257), the chromatic adrenergic neurons have a distribution which is similar to that observed by Ballowitz (254) by means of the silver impregnation technique.

Synaptic contact by which unmyelinated nerve fibers form a "chromatophore-neural junction" with dermal melanophores has been observed (259, 260). A single electrical stimulus to nerves causes a submaximal melanophore contraction *in vitro* (261). Since the response was not "all or none," this suggested that two or more neurons control each melanophore. The demonstration of an antidromic response indicated that some nerves apparently innervate more than a single melanophore. Since tetrodotoxin blocks the response of melanophores to electrical stimulation without affecting melanosome movements or their subsequent ability to contract in response to catecholamines, this result was taken (261) as evidence that melanophore responses are due entirely to nervous excitation rather than to a generalized spread of electrical current to the pigment cells.

It has long been claimed that teleost melanophores are doubly innervated, a view strongly advocated by Parker (1, 115). Parker favored the view that melanophore contraction is mediated through a sympathetic (adrenergic) event, whereas melanosome dispersion is controlled by the parasympathetic system, acetylcholine being the neurotransmitter. Although numerous investigations on a great number of teleosts support an adrenergic regulation of melanophore contraction, there is no recent evidence in support of a double innervation theory, nor of a parasympathetic involvement in melanosome dispersion (262, 263). If the melanophore is regulated solely through a contracting nerve fiber, then dispersion may apparently result from any interruption of the influence of these fibers by any means (263). Parker felt that dispersion was an active process (1).

The early investigations of Hogben and Mirvish (264) clearly delineated a role for the central nervous system and a sympathetic regulation of melanophore contraction in the chameleon (*Chameleo pumilus*). According to Hogben and Mirvish, there is no evidence for a hormonal control of chromatophores in chameleons, but as Parker (247) has suggested, a role for intermedin was not entirely ruled out. Canella (122) has demonstrated in another chameleon (*C. jacksoni*) that both MSH and ACTH can cause darkening through melanosome dispersion. In *Phrynosoma,* both nerves and hormonal agents (247, 265, 266) appear to play a role in chromatophore control. In *Anolis carolinensis,* in contrast, the careful studies of Kleinholz (117, 118, 267) clearly demonstrated that only hormones, and not nerves, are involved in chromatic control in this lizard. So few studies have been undertaken on color control in reptiles that the present extent of our knowledge of the subject is still mainly based on the earlier investigations of *Chameleo, Phrynosoma,* and *Anolis.*

The adaptive chromatic responses of elasmobranchs are very slow and consistent with a humoral theory of chromatophore control. Wykes (268) could find no evidence for a direct innervation of melanophores in elasmobranchs (*Scyllium canicula, Raia brachyura, R. maculata,* and *Rhina squatina*), thus confirming the earlier investigations of Young (269), who reported the absence of sympathetic innervation of the skin and chromatophores of seven species of elasmobranchs. Both Parker (270) and Vilter (139) suggested a direct nervous innervation and control of elasmobranch melanophores, but Abramowitz (271) has criticized these nervous theories and found them unconvincing and wanting in completeness.

There is no evidence for a nervous regulation of chromatophores in the Amphibia (272). Although there is strong evidence that the presence or absence of circulating intermedin can entirely account for adaptive background responses, it should be considered a possibility that certain chromatophores making up the stripes or spots in some anurans, such as the leopard frog, *Rana pipiens,* might be innervated and, therefore, separately controlled. Frogs do not "lose" their spots when they adapt to a light-colored background. It is during such a response that the spots stand out and provide the frog with a disruptive pattern. Contrary to common belief, these spots are not areas of the integument where there are necessarily greater numbers of melanophores; rather, the spots are composed of melanophores that appear to be under a different physiological regulation. The melanophores of the spots remain expanded while the melanophores between the spots become contracted in response to a white background and hence the absence of intermedin. It is quite possible that the spot melanophores are innervated and remain expanded, in the absence of intermedin, through direct nervous stimulation. These spot melanophores are, however, physiologically responsive, as attested by the fact that they contract when the skin is removed from the frog and is placed in Ringer solution. These dermal spot melanophores are not physiologically equivalent to dermal interspot melanophores, for they show differential responses to hormonal and neurohumoral agents (36).

Studies of normal human skin, of pigmented nevi, and of melanomas (273, 274) by the histofluorometric technique have all failed to demonstrate adrenergic innervation of melanophores. Connections between cholinesterase-positive nerve fibers originating in the dermis and epidermal melanocytes have been noted in the human fetus (275). Certain depigmented conditions in man, however, appear symmetrically bilateral in nature (Fig. 5–4); they have been considered as perhaps deriving from alterations in melanophore function as a consequence of altered hormonal stimuli, possibly of nervous origin (276). According to Lerner, hypopigmented areas of the skin are frequently confined to skin regions supplied by a nerve segment. The observation of increased sweating in such regions indicated the occurrence of some type of increased nervous activity in these vitiligous areas (277). Lerner concluded that vitiligo might result from increased activity of peripheral nerve endings.

PHARMACOLOGICAL CONSIDERATIONS

Effects of
neurotransmitter agents on chromatophores

Since all vertebrate chromatophores are of a similar ectodermal origin, as are peripheral nerves, they might be expected to respond to the two common vertebrate neurotransmitters, norepinephrine and acetylcholine. In view of the fact that a sympathetic innervation of melanophores has been demonstrated for some teleost fishes, as well as reptiles, it is not surprising, therefore, that the catecholamines—norepinephrine and epinephrine, which are structurally quite similar—affect the movement of pigment organelles within chromatophores of many poikilotherms.

In vitro studies on cyclostome (123), teleost (278), amphibian (279), and reptile (*Phrynosoma,* 265, 266; *Anolis,* 118) melanophores have demonstrated the direct contracting effects of catecholamines on these pigment cells. These same catecholamines can also cause the opposite response—

melanophore expansion—in teleosts (255), amphibians (280, 281), and reptiles (118, 119). The ability of catecholamines to evoke either a so-called excitement pallor or excitement-darkening, which results from melanosome aggregation or dispersion, depends on the nature of the adrenergic receptors possessed by the melanophores, as discussed in the next chapter.

Since the vertebrate adrenal medulla may release either or both epinephrine and norepinephrine—depending, among other things, on the species—under conditions of stress, these catecholamines may certainly determine the chromatic responses that are often observable under such conditions of excitement. In addition, injected catecholamines are known to affect the release of a number of hypothalamic releasing or inhibiting factors that regulate pituitary function. Moreover, since an adrenergic regulation of pars intermedia control has been suggested, as discussed earlier, it is quite possible that excitement pallor may be reinforced through a hypothalamic inhibition of MSH release, in addition to the direct effects of adrenal catecholamines at the effector cell (chromatophore) level. Such a suggestion is compatible with the presently uninterpreted results of the effects of injected catecholamines on the African clawed toad *Xenopus laevis* (280). Epinephrine injected into this anuran causes a drop in melanophore index of black-adapted toads from 5 to 3 (282). Similar injections of this catecholamine into white-adapted or hypophysectomized individuals cause a rise in melanophore index from 1 to 3, and the same results take place *in vitro*. The inhibition of pars intermedia release of MSH concomitant with a direct, but submaximal, stimulation of melanosome dispersion adequately explains these apparently anomalous observations.

Both norepinephrine and epinephrine can induce either a dispersion or an aggregation of reflecting platelets within amphibian iridophores (283). The effects of catecholamines on the bright-colored chromatophores of other poikilothermic vertebrates has not been studied in enough depth to warrant any further discussion.

Acetylcholine is the neutrotransmitter of the postsynaptic neurons of the parasympathetic division of the autonomic nervous system. It is also released from the preganglionic neurons of both the parasympathetic and sympathetic divisions of the autonomic nervous system. In addition, its presence within the central nervous system, including the hypothalamus, suggests a possible integrative role at that level. It is not surprising, therefore, that injections of acetylcholine may result in a number of possible physiological effects on chromatophores, in addition to direct effects. Since a cholinergic link inhibitory to an adrenergic inhibition of MSH release has been indicated in pars intermedia control (158), injections of acetylcholine may darken poikilotherms by releasing intermedin. The chromatic responses induced by acetylcholine are complicated by the fact that this compound may stimulate ganglionic transmission, thereby resulting in norepinephrine release from postsynaptic neurons innervating melanophores in those vertebrate melanophores so regulated. Furthermore, acetylcholine normally is responsible for the release of adrenal catecholamines, and injections of the former may cause lightening of animals through stimulation and release of either epinephrine or norepinephrine. Such possible effects, if not considered, can certainly lead to misinterpretations relative to mechanisms of chromatic regulation.

To complicate matters further, the chromatophores (dermal and epidermal melanophores, as well as iridophores) of the frog *Rana pipiens* possess cholinergic receptors, and acetylcholine causes a profound *in vitro* lightening of MSH-darkened skins. This point clearly demonstrates the direct effects of this neurotransmitter at the effector cell level. Only about one out of three frogs of this species of anuran possesses chromatophores sensitive to acetylcholine (284), thus again further complicating interpretations, even within a single species of frog.

Although Parker (1) tried to implicate both the parasympathetic as well as the sympathetic nervous system in the regulation of teleost chromatophores wherein acetylcholine would control melanophore expansion, all subsequent investigations have failed to substantiate such a control. If acetylcholine has a direct effect on teleost melanophores, then, as Parker himself first demonstrated (285) but later tried to discredit (286), it causes melanophore contraction *in vitro*, as in amphibians. Acetylcholine is without direct effect, except possibly at very high concentration, on the chromatophores of the few reptilian species (e.g., *Anolis*) that have been studied (99).

ROLE OF STEROIDS
IN CHROMATOPHORE CONTROL

Adrenal steroids have generally not been implicated in the direct control of chromatophores. As discussed earlier, their indirect effects on either or both ACTH and MSH release may have profound effects on integumental pigmentation. Cortisol has been reported to reverse the action of MSH of frog skins *in vitro* (287) and *in vivo* (288). Other workers, however, have been unsuccessful at similar *in vivo* attempts (289). The *in vitro* lightening effects of cortisol (succinate) are quite minimal, and the possibility that the succinate moiety itself is partly responsible for the lightening effect has not been ruled out (288).

Steroidal hormones of gonadal origin, however, have been implicated in directly modifying skin coloration in mammals. A number of investigations indicate that estrogens, whether applied locally or injected systemically, are capable of stimulating melanogenesis in some mammals. After ovariectomy in the guinea pig, the melanin content of epidermal melanophores is reduced and thus there is a decrease in the visible length of the dendritic processes of the melanophores (290). There is also a reduction in melanophore numbers in the areolae, accompanied by a decrease in the amount of cytocrine melanin. Estradiol administered to female guinea pigs causes increased melanogenesis within the skin, particularly within melanophores of the areolae (3, 291). The effects of the estrogen are direct, for local application of estradiol to one nipple causes a localized increase in melanin formation without stimulating the other nipples. Darkening of the nipples and areolae of humans accompanies pregnancy; and as the period of gestation increases, the areolae become darker and wider. The areolae decrease both in size and degree of pigmentation after parturition. These normal physiological changes, therefore, correlate well with experimental observations and suggest a normal physiological role for estrogens in mammalian pigmentation. Whether the so-called direct effects of estradiol are synergistic with circulating MSH is unknown.

In the male mammal, testosterone may also be responsible for localized pigmentogenic effects. Castration of the male mouse maintains the glossy black "juvenile" coat in certain genetic strains, while intact controls molt to a normally duller coat color in which individual hairs contain less melanin (292). The dimorphic pigmentary pattern between the sexes of the golden hamster is under control by testosterone (293). The male normally has a dark pigmented area of skin in the dorsolateral subcostal region, whereas female hamsters may either show no pigmentation at all in the area or only a slight amount. This area in both sexes becomes darker in response to injected testosterone, but not estradiol. Again, this pigmented area undergoes normal changes in coloration that correlate well with normal endocrine changes in testicular release of testosterone. Similarly, in the black strain of Long-Evans rats, injected testosterone stimulates melanin production within epidermal melanophores of scrotal skin (294, 295).

The selective anatomical effects of sex hormones can be explained by either of two mechanisms (3). The melanophores in certain anatomical regions may be (1) more sensitive to these hormones than others situated elsewhere, or (2), as has been suggested as more plausible, the hormones stimulate the surrounding epidermal cells in these areas of hyperpigmentation to become selectively more active in taking up melanin granules from melanophores.

The effects of progesterone on pigmentation are too varied in mammals to make any clear evaluations. Progesterone is, however, a potent darkening agent of frog skins, *in vitro* (288, 296). It causes melanosome dispersal within both dermal and epidermal melanophores, as well as re-

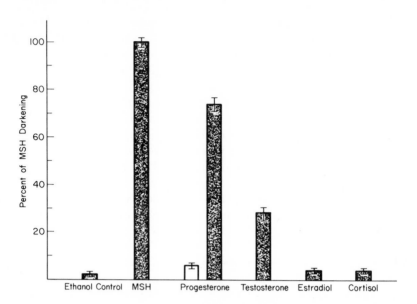

FIG. 6-5. *In vitro* response of frog (*R. pipiens*) skins to various steroids: progesterone (10^{-5} *M*, open bar; 10^{-4} *M*, solid bar); testosterone (10^{-4} *M*); estradiol benzoate (10^{-4} *M*); and cortisol acetate (10^{-4} *M*). The darkening response is expressed as a percentage of the maximal darkening response elicited by a high concentration (4×10^{-9} g/ml) of MSH. The responses are also compared to an ethanol control group of skins. Each bar is the average reflectance value of eight skins after 2-hours treatment. Vertical lines indicate the standard error of the means. [Redrawn from (288)]

flecting platelet aggregation within iridophores, thus mimicking the action of MSH. Testosterone is only weakly effective in darkening the skin, and estradiol and hydrocortisone are totally without such an effect (Fig. 6–5). It is difficult to assess the effects of gonadal hormones in mammals *in vivo* because of their possible synergistic effects with other hormones. The *in vitro* effects of progesterone on frog melanophores, as previously observed (288, 296), is a direct one and takes place in the absence of MSH or other hormones. This steroid directly disperses melanosomes within epidermal melanophores, whose functions in the frog are similar to those in mammals, and suggests a possible melanogenic role in the vertebrate epidermis.

In birds, both estrogens and androgens can have profound effects on plumage coloration in certain species. The only pigment cells within the feather follicles are epidermal melanophores, and gonadal steroids affect the type of melanin synthesized within these chromatophores. Sex hormones may also influence the deposition of pigments in the beak of some birds (Fig. 2–5). The effects of these steroids are local and appear to direct the pathways of melanin synthesis to either eumelanin or the phaeomelanins. The result, therefore, is the formation of either the dark melanins or the lighter red or yellow melanins. Plumage patterns within an individual bird may depend on the preferential synthesis of a particular type of melanin within specific anatomical areas. Hormonal effects on follicular melanoblasts are apparently mediated through the cellular substrate surrounding these cells rather than directly on the cells themselves (39). The varied response of melanophores of the same color pattern genotype can be "attributed to regional differences in the physiological properties of the feather papillae themselves," and, further, "these properties vary from tract to tract and from feather to feather within the tracts" (297). A number of avian species have been studied, but the endocrine or other mechanisms controlling plumage changes in most species is still little understood (38). The precise interrelationships of pituitary gonadotropins, gonadal steroids, and genetic constitution on feather coloration within an individual species of bird are not entirely clear. Luteinizing hormone causes melanization of feathers in African weaver birds (298, 299) even after castration (300). The locus of action of the extragonadal effect of luteinizing hormone is unknown, and presently no evidence exists for a local action of the hormone on the regenerating feather (301).

In a number of species of both teleosts and reptiles, there may be a seasonal assumption of nuptial coloration, as well as permanent sexual dimorphic differences in pigmentation. In teleosts, at least, these differences can be mimicked by gonadal steroids (302). Whether the effects are direct or not is unknown. Little information is available concerning the mechanisms of control of nuptial coloration in reptiles. Dimorphic differences in nuptial coloration in poikilotherms result from the presence or absence and response of dermal xanthophores or erythrophores, whereas in mammals and birds such differences in nuptial coloration relate to the nature of epidermal melanin.

ROLE OF THE THYROID
IN CHROMATOPHORE CONTROL

Although seldom discussed as an important factor in the regulation of integumentary coloration, there is evidence that thyroid hormones may influence pigmentary events, particularly at certain developmental stages. In amphibians, iridophores become especially abundant in the skin during metamorphosis. The white spots of some adult ambystomid salamanders develop as a result of iridophore proliferation. Woronzowa (303) showed that this spotting is controlled by hormones. In salamanders injected with thyroid extracts or kept in suspensions of thyroid glands, there appears an increase in the amount of the surface of the animal covered with white spots. It was concluded that the spots of these salamanders occur as a result of the hyperthyroidism existing during metamorphosis. Although it is not known whether these effects of thyroid extracts are direct or not, the development of white spots during metamorphosis could be retarded and the spots finally made to disappear by pituitary injections.

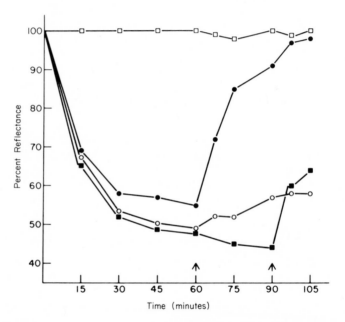

FIG. 6-6. Effect of triiodothyronine on MSH-darkened skins of *R. pipiens*. Three groups of skins were darkened with porcine–MSH (0.75 × 10^{-9} g/ml) for 60 minutes. Triiodothyronine was then added at a 10^{-5} *M* concentration to one darkened group (●) and to a Ringer control group (□) of skins. Triiodothyronine at 10^{-6} *M* (○) and norepinephrine (■ , 10^{-5} *M*; at 90 minutes) were then added to the other two MSH-darkened groups of skins. Each point on the graph is the mean of eight measurements of reflectance. [From (21)]

Thyroxine may affect the circulating level of MSH in the frog by a hypothalamic or pituitary inhibition of MSH release from the pars intermedia (304). In addition, both thyroxine and triiodothyronine have a direct antagonistic effect on MSH darkening of frog skins *in vitro* (305). Triiodothyronine is more potent than thyroxine in reversing the action of MSH on isolated frog skins. In addition, triiodothyronine is more effective than any other known hormonal agent *in vitro* in lightening MSH-darkened frog skins (Fig. 6–6). The reason is that triiodothyronine reverses the action of MSH on iridophores, as well as on epidermal and dermal melanophores (21). The lightening of frog skins by other hormonal agents is, in contrast, much less effective, for these agents have differential effects on chromatophores and not all skins from any one or more animals respond. Both thyroxine and triiodothyronine totally reverse the darkening action of MSH on lizard (*Anolis*) skins. No other hormones are as effective *in vitro* on both amphibian and reptilian skins.

Chronic or permanent pallor resulting from melanosome aggregation, followed by reduction in integumental melanophore numbers in salmonids, is a consequence of induced hyperthyroidism (306). A silvering, resulting from integumental deposition of guanine, is elicited by thyroxine, thyroid extracts, or thyrotropin (307). Conversely; hypothyroidism resulting from radiothyroidectomy induces a gradual but permanent darkening of young rainbow trout that results from an increase in the number of integumental melanophores (308). Melanophores come to occupy areas of the skin that are normally only sparsely pigmented (Fig. 6–7). These morphological color changes involving increases or decreases in melanophore and iridophore numbers in the skin, as experimentally induced by administration of thyroid hormones, may possibly have their natural counterpart in the color changes

FIG. 6-7. Pigmentary differences between a control (top) and a radiothyroidectomized trout. [From (308)]

that take place during smoltification in preparation for seaward migrations, when thyroid activity apparently increases. A lack of an effect on pigmentary change has been reported for some species of fish following injections of thyroid preparations, thyroxine, or thyrotropin (302).

In birds, thyroxine has contrasting effects among different species. The amount of melanin pigment can be either decreased or increased. As is true for steroid hormones, the nature of epidermal melanin synthesized within melanophores is apparently affected by thyroid activity, thus leading to the formation of differently colored melanins.

CHAPTER
7

chromatophore control III

*control of
invertebrate color changes*

Chromatic adaptations of invertebrates, as of vertebrates, result from pigment organelle movements within integumental chromatophores and are similarly regulated by either humoral or nervous innervation. In addition, as for vertebrates, both rapid, physiological color changes and slower, morphological color changes are involved in chromatic adaptations. Although the number of invertebrate species far outnumber that of vertebrates, detailed studies on the structural, functional, and regulatory basis for color change is limited to a few species of cephalopods and crustaceans. We must here limit our discussion to those few species of invertebrates that have been investigated in some depth. A number of reviews on the subject have been published (1, 309–312).

CONTROL OF CEPHALOPOD CHROMATOPHORES

The rapid and colorful chromatic displays of cephalopod mollusks have attracted the attention of many naturalists, including Charles Darwin. While on his voyage in the *Beagle,* he described (313) his interest in the "very extraordinary, chameleonlike power of changing their colour" of the octopus, and he noted that they "vary their tints according to the nature of the ground over which they pass. . . ." Actually, both the cytological and physiological features of cephalopod color change were described even before the existence of chromatophores in other animals had been discovered (314). However, only recently has the definitive structure (30) and physiological regulation (315) of cephalopod chromatophores been clearly described.

Cephalopod pigment cells differ from other invertebrate, as well as vertebrate, chromatophores in that they are actually organs rather than autonomously functioning individual cells. These chromatophores may be brown, black, red, or yellow. The movements of pigments within these chromatophores are regulated by muscle fibers that are attached to the pigment cell proper (Figs. 2–16 and 2–17). The response—contraction or relaxation—of the attached muscles themselves regulates the shape of the pigment cell and hence color change. Chromatophore control is therefore regulated indirectly through nervous innervation of chromatophore muscles rather than directly as in other animals.

Chromatophore organs are entirely under central nervous system control. Neurotransmitter release from nerve endings leads to muscle contraction, thus resulting in melanophore expansion. Conversely, in the absence of neurohumoral stimulation, muscles are relaxed, thereby allowing for melanophore contraction. Each chromatophore is innervated by more than one neuron (316); each motor axon may activate several muscle fibers of an individual pigment cell (314). Selective activation of individual motor fibers permits a delicate control of integumental coloration via a step-by-step expansion of the chromatophores as more muscles are recruited through polyneuronal nervous stimulation. There is no morphological evidence in *Loligo opalescens* for the existence of more than one type of nerve axon. These axonal terminals do not seem to be monoaminergic in nature. In *Loligo vulgaris,* on the other hand, more than one type of neuron may be present (317). Although both acetylcholine and serotonin may have effects on cephalopod chromatophores, neither neurotransmitter agent has been clearly implicated in the normal control of chromatophores. The nature of the transmitter substance of the motor axons is therefore yet to be established. All evidence indicates that chromatophore muscle fibers are innervated by motor fibers only; there is no evidence for an inhibitory innervation (315).

Comparative studies of similar elegance to those described by Florey and his co-investigators for the chromatophores of *Loligo opalescens* would be a valuable contribution to pigment cell physiology. Cloney and Florey (30) have suggested a number of diverse and interesting problems that are now in need of investigation.

CONTROL OF CRUSTACEAN CHROMATOPHORES

In contrast to cephalopods, crustacean chromatophores are controlled through humoral rather than nervous agencies; but, like vertebrates, some chromatophores may respond directly to photic stimulation, and the degree of response may be modulated by temperature effects. Crustacean chromatophores may be black, yellow, red, or white and may be intimately associated with one another to produce a polychromatic effect. Early investigations on crustacean chromatophore systems were concerned with establishing the nature of chromatic adaptations of various species and the possible humoral basis for such color changes. A great impetus to such studies followed the discovery that the humoral agents were of nervous origin and apparently neurosecretory in nature. Most investigators are presently concerned with determining the chemical nature as well as the number of chromatophorotropins involved in chromatophore control.

Some crustaceans (e.g., the prawn *Palaemonetes*) chromatically adapt in conformity with the nature of the background while others (e.g., fiddler crab, *Uca*) do not. *Uca* does, however, exhibit daily rhythms of color change unrelated to the nature of the color of the background. Some crustaceans (e.g., *Hippolyte*) both background adapt and exhibit diurnal rhythms of color change. Since rhythms of chromatic activity are abolished by eyestalk removal, a periodicity in release of a chromatophorotropin is suggested. Because *Uca* will continue to exhibit diurnal color changes for many days while in complete darkness, it would appear that these rhythmic chromatic activities are to a great extent independent of background or lighting conditions (1). A tidal rhythm of color change paralleling the 12.4-hour ocean tides has a modulating effect on the daily rhythm to produce semimonthly patterns of fluctuation in the daily color change cycles. The tidal rhythms of color change of a particular fiddler crab are apparently adjusted to tidal times of their own local habitat tidal changes (318).

Pouchet (319) discovered that removal of the eyestalks from certain crustaceans led to a loss of adaptive chromatic responses. He considered that the eyes, by way of nervous elements, regulated chromatophore activity, although he was unable to inhibit chromatophore responses by sectioning various nerves. Perkins (320) determined that the chromatophores of *Palaemonetes* were aneuronic, and he actually provided data that the controlling agency might be of humoral origin. This point was substantiated when Koller (321) observed that blood taken from a dark *Crangon* injected into a pale one caused darkening of the recipient, even when it was kept on a white background. Eyestalk extracts from pale *Palaemonetes* injected into blinded *Palaemonetes* (which are dark in color due to blinding) caused chromatophore contraction. Other tissue extracts or saline solution did not induce such a color change. The role of the eyestalk in the regulation of crustacean chromatic responses has been established for numerous other species. Eyestalk extracts from one species were found to be equally effective in some other species.

The source of the crustacean eyestalk chromatophorotropin(s) has been shown to be the X-organ-sinus gland complex, a neurosecretory organ. Other elements both within and outside the eyestalk are also apparently involved in chromatophore control (322). Under conditions of constant illumination, blinded *Palaemonetes* remain dark, whereas blinded *Uca* remain pale. This result would seem to indicate, therefore, as early pointed out by Abramowitz (323), that if other tissues capable of forming chromatophore hormones are present, "they play an insignificant part in the ordinary chromatic physiology" of crustaceans.

Brown (324), however, did provide evidence that chromatophorotropins might exist in nervous system organs other than the eyestalk. This possibility is suggested by the observation of the persistence of some degree of diurnal darkening in eyestalkless *Uca* (325). Chromatophorotropic substances are apparently also present in supraesophageal and thoracic ganglia as well as in the circumesophageal connectives (326).

Although the loss of eyestalks in *Palaemonetes* and *Crangon* leads to permanent darkening, similar ablation in the fiddler crab, *Uca,* results in blanching. A startling observation was that extracts from *Uca,* which darken this crab itself, cause lightening of *Palaemonetes*. Similar extracts from *Palaemonetes* that lighten this shrimp will darken *Uca*. These results suggested that crustacean chromatophores themselves differ radically between species (1). Indeed, similar differences in what might be referred to as the "resting state" of chromatophores exist between other natantian (shrimps; e.g., *Palaemonetes*) and brachyuran (true crabs; e.g., *Uca*) decapods.

Kleinholz has suggested (327) that the crustacean chromatophorotropins are relatively small peptides. But as he has pointed out (328), "the important question of how many distinct chemical entities are involved as regulatory hormones in crustaceans" is largely unknown. Eyestalk extracts of the prawn, *Pandalus jordani,* have been chromatographed. Four different biologically active peaks were obtained. The substance in each peak was effective in causing the light adaptation response of distal retinal pigment (see later discussion) as well as in dispersing the pigment in the black, red, and white pigment cells. It was suggested (328) that these "pigmentary effector hormones may be closely related but slightly different molecules, or they may be a single molecule acting on the four different pigmentary effectors."

According to Fingerman (311), pigment-concentrating as well as pigment-dispersing substances have been described for the four color varieties of chromatophores of *Uca pugilator*. But as similarly suggested above by Kleinholz, it is not known whether each type of chromatophore is controlled by a different chromatophore-concentrating and -dispersing substance or whether a single chromatophorotropin may activate more than one pigment cell type. Both possibilities might obtain, for, as pointed out for vertebrates (e.g., the frog, *Rana pipiens*), a single hormone (intermedin) contracts iridophores but disperses melanophores and xanthophores. In addition, catecholamines can either aggregate or disperse pigment organelles within both melanophores and iridophores, and the particular chromatophore response may differ between species or even anatomically within an individual. Thus a single hormone in crustaceans could conceivably also control a number of diverse pigmentary events.

It would appear that most workers favor a polyhumoral theory of chromatophore control rather than a unihumoral or even bihumoral concept. Those following the multiple hormone theory refer to the crustacean chromatophorotropins as the red pigment-dispersing hormone (RPDH) and the red pigment-concentrating hormone (RPCH) ; similar abbreviations are used to designate the white (WPDH, WPCH) and black (BPDH, BPCH) pigment-dispersing and -concentrating hormones. Until their chemical identities can be ascertained, caution should be exercised in any discussion of their supposed separate activities. *Palaemonetes* can adapt its color to backgrounds of white, yellow, red, green, blue, dark-grey, and black (329). This suggests a complex mechanism of control, for each of the variously colored chromatophores responds differently.

A possible problem in the study of crustacean chromatophore control concerns the antagonisms that may exist between each of the postulated red, black, or white pigment-concentrating and -dispersing hormones. Results obtained by injecting extracts of nervous tissue may reveal only the net effect of the relative amounts and potencies of the various chromatophorotropins that might be present in the extracts (330). The sex, age, and state of the intermolt period, as well as the reproductive state of the individual, may influence the response of pigment cells to chromatophorotropins (331).

A further complicating factor in the *in vivo* actions of injected chromatophorotropins is that eyestalks may contain a substance which causes chromatophorotropin release from nervous tissue (332). The isolation and further study of this postulated releasing factor would be interesting in light of the existence of neural factors in the regulation of intermedin release in vertebrates. As in vertebrates, biogenic amines (possibly serotonin) may play a role in the release of crustacean chromatophorotropins (333). The observation that the injection of serotonin indirectly results in dispersion of erythrophore pigments, but not that of melanophores, might be taken as evidence that RPDH and BPDH are separate hormones (334).

A most interesting mechanism of color control has been studied (335) in the marine isopod, *Idothea montereyensis*. This crustacean lives on marine plants of rocky coasts, and distinct color varieties match the various colors of the plants on which the species is found. Individuals may be red, green, or brown. These isopods have the ability to change color and match that of the plant to which they are transferred. These colors are mainly the result of cuticle pigmentation which is partially modified or enhanced by the colored chromatophores that are present. It is interesting, therefore, that although these isopod crustaceans lack a diversity of chromatophore types, they can assume a number of color phases by an interaction of cuticular and chromatophore pigments. Cuticle pigmentation changes occur only after an animal has completed a molt. The two plant carotenoids that are most abundant in the plant materials consumed by this isopod are "selectively utilized and stored by these isopods" (336). These animals need light in order to accomplish a color change, and this environmental cue apparently dictates the quality and quantity of pigments deposited in the cuticle. The red chromatophores occur in specific states of contraction or expansion in each of the three color varieties and reinforce cuticle coloration. It would be interesting to determine whether chromatophorotropic agents regulate both the biosynthetic pathways of the particular cuticular pigments synthesized as well as the chromatophore responses.

In addition to integumental chromatophores, pigments are also found within the ommatidia, the photoreceptive units of the compound eye of crustaceans. These complex structures are composed of a number of specialized cell types and possess both distal and proximal black pigments as well as a reflecting pigment. Although the details of the movements of the three pigments have been clearly described for *Palaemonetes* (337), a cytological study of the specific cellular localizations of these pigments within the ommatidia of this shrimp, as well as in other crustaceans, is in need of clarification at the ultrastructural level. It is not known whether the three pigments are separately localized within individual chromatophores or whether two or more of the pigments reside within a single cell. The movements of the pigments are in response to the level of incident illumination, but the positional changes that the retinal pigments undergo vary among species (338). The positional movements of the pigments function to regulate the amount of light to which the photosensitive cells of the ommatidia are exposed.

There is good evidence for an eyestalk hormone controlling the position of the distal black pigment in the light-adapted state. Removal of eyestalks results in a shift in position of the distal black pigment, which can be inhibited or reversed by injections of eyestalk extracts. There is only limited evidence for a hormonal regulation of the proximal black pigment and the reflecting pigment under conditions of either light or dark adaptation.

Persistent rhythmic movements of retinal pigments in crustaceans under conditions of constant illumination (339) or darkness (339, 340) have been described. Kleinholz (337) early asked whether these different rhythms might be attributable to several different hormones or to one hormone to which the various pigment cells respond, depending on their threshold differences. In addition, are these migrations the results of liberation of a hormone or of the cessation of hormone secretion? These questions remain unanswered. It is quite possible, however, that diurnal retinal pigment movements may depend on a periodic hypoxia of the cell systems concerned (341).

It is interesting that, in the vertebrate eye, pigments also play an important role in ocular function. Both melanophores and reflecting cells may be present within the eye of most vertebrates. The migration of melanosomes within retinal or choroidal melanophores controls the amount of light that reaches or is reflected to the retinal photoreceptors. In darkness, melanosomes within the pigmented cells surrounding the rods and cones move to the back of the retina; in light, these melanosomes disperse through dendritic processes that penetrate to the photoreceptor layers of the retina. The behavior of the retinal pigment migrations over a wide range of illuminations has been studied in salmon (342). As in crustaceans, pigment granule movements within the eye may be hormonally regulated (343, 344).

CONTROL OF
COLOR CHANGES IN INSECTS

The color of some insects matches the color of their surroundings, often in a very striking manner. For example, four adaptive color responses occur in the migratory locust *Oedipoda coerulescens*—namely, red (on iron oxides), black (on shale), grey (on Fontainbleau sand), and yellow (on yellow clayey sand) (345). It is most probable that these physiological color changes are regulated through a visual mechanism. In the Alpine grasshopper, *Kosciuscola tristis,* on the other hand, physiological color change is apparently a thermoregulatory response (346). Males are a bright greenish-blue above about 25° C and a dull, almost black color below about 15° C, and intermediate shades develop at in-between temperatures.

The cytological basis for color change in *K. tristis* has been clearly described by Key and Day (346). A dense layer of small and highly refractive granules occupies the distal outer portions of the hypodermal cells of the integument, and it is underlain by a layer of larger and darker granules within the same cells. In the dark phase of this hopper, the position of these granular layers is reversed, and at intermediate color shades the granules show transitional distributions. Movements of intracellular granules within hypodermal cells of the stick insect, *Carausius morosus,* have been observed (347). Whether these cellular responses can be considered as chromatophore-like in nature and a common feature of the insect integument is unclear. Other chromatophorelike responses have been described for the black pigment-bearing cells of the tracheal air sacs of the larvae of the dipteran, *Corethra* (348).

The physiological color changes of the stick insect, *Carausius morosus,* were early studied in great detail (see 1, for references). Both moisture and environmental illumination affect color change. Both responses to darkness and to moisture result in darkening of the insect. These responses, although received through separate sensory receptors, each involves the release of a blood-borne hormonal agent liberated from the brain by stimulation of afferent nervous pathways. The eye is the initial sensory receptor regulating color change in response to environmental lightening, and continued stimulation leads to morphological color change (312).

Clearly, then, hormonal mechanisms are involved in color control in some species of insects. Recent studies have implicated the hormones ecdysone and bursicon as playing specific roles in the cuticular coloration (tanning) during larval and adult development of such blowflies as *Calliphora* and *Sarcophaga.* In the puparium of *Calliphora,* N-acetyldopamine has been identified as the tanning agent; its control by ecdysone involves induction of the key enzyme, dihydroxyphenylalamine (DOPA) decarboxylase (349). In the adult, on the other hand, the mechanism of action of the tanning hormone, bursicon (350, 351), is different from that involved in tanning of the puparium. Bursicon does not initiate the *de novo* synthesis of any of the enzymes required for the tanning process. It appears, rather, that bursicon acts by activating tyrosine hydroxylation (352).

Bursicon appears in the blood of newly emerged flies, perhaps in other insects as well. The blood of earlier developmental stages is devoid of the hormone. It has been found in the blood of insects of four other orders: Orthoptera, Hemiptera, Coleoptera, and Lepidoptera. Bursicon is apparently elaborated by neurosecretory cells of the insect brain and then transported to thoracic ganglia. Its release is controlled by nervous stimulation from the brain, which is itself activated by stimuli received from the cuticle after the fly emerges from the substrate in which it was previously confined. In the American cockroach, *Periplaneta americana,* bursicon apparently acts directly on the epidermis, for isolated pieces of cuticle with attached epidermis when incubated in Ringer solution with the hormone become tanned (353). Fraenkel and his colleagues have shown that bursicon is not one of the previously described insect hormones (351). Although both ecdysone and bursicon regulate these morphological color changes in dipterans, and are apparently present in a number of other insect orders, it is not known whether these hormones also regulate the physiological color changes described earlier for some insect species (such as *Corethra*).

Many grasshopper species are polymorphic with respect to color and other morphological features. The phase theory of Uvarov (354) was formulated to explain the interrelationships of the different forms of *Locustana pardalina*. The phase theory postulates that the species exhibits the characters of the phase *gregaria* when it forms large swarms and that individuals living in isolation cannot retain these characters but must acquire those of the phase *solitaria* (355). Grasshoppers in the *solitaria* phase are a green or buff color and in the extreme *gregaria* phase are almost totally black. It has been shown that there is an external, airborne factor, apparently a pheromone, affecting the degree of deposition of the dark cuticular pigment (356).

Seasonal color polymorphism of the butterfly *Colias eurytheme* appears to have evolved to maximize solar heating in cold seasons and to minimize overheating in warm seasons (357). Specimens of the species *Colias* from cold climates have much darker and more heavily melanized undersides than do individuals from warm climates. Reproductive success in cold climates might well depend on the efficiency of absorption of solar energy for heating and thus increased activity. Light-colored *Colias* from warmer habitats, on the other hand, might be less susceptible than dark individuals to heat stress and a possible forced inactivity (358).

Industrial melanism in moths is controlled genetically and is brought about by the indirect effects of industrialization through natural selection (359). Experiments on the peppered moth *Briston betularia* suggest that there is cryptic advantage of the light form (f. *typica*) in unpolluted and lichened woodland and of the darker form (f. *carbonaria*) in the industrial countryside.

COLOR CONTROL
IN SOME OTHER INVERTEBRATES

Annelids

The occurrence of chromatophores in certain annelid worms was early noted by Claparède (360). He observed that the pigment within the yellow and the violet chromatophores of *Nereis dumerilii* was either in the aggregated or dispersed state. Numerous workers (see 361 for references) have noted that chromatophores of other annelids (*Protoclepsis tessellata, P. geometra, Hemiclepsis marginata,* and *Glossosiphonia complanata*) become contracted in the dark and dispersed under conditions of illumination. Color change in the leech, *Placobdella parasitica,* are due mainly to green chromatophores that are punctate under conditions of darkness but that become dispersed when the worm is exposed to light. It was clearly demonstrated (361) that the dispersion of pigment is brought about by nervous action mediated through the central nervous system. Contrary to an earlier suggestion (362, for *Glossosiphonia complanata*), there is no evidence of a direct effect of light on the chromatophores. The smaller red pigment cells that are also present in *P. parasitica* react independently of the green ones, but it appears that their pigment, too, is concentrated by a central nervous mechanism. These results implying a nervous control of chromatophores are in need of further study to determine the neurohumoral transmitters involved in pigment granule movements.

Echinoderms

By day, or in strong light, young echinoderms, *Diadema antillarum,* are almost uniformly black. They lighten and develop a striking white pattern at night or when placed in the dark (363). These color changes result from melanin pigment movements within melanophores. A narrow beam of light projected on the test causes a localized darkening in color, a response that is direct, for it occurs in small isolated pieces of test. When a part of an individual chromatophore of *Diadema setosum* is illuminated, the pigment remains dispersed. If, however, a pigment-free area of a chromatophore is illuminated, the cell disperses the pigment so as to cover the illuminated area (364). It was concluded, therefore, that the photosensitivity is a property of the cytoplasm rather than of the pigment granules.

Von Uexküll (365) had earlier observed that the echinoids, *Arbacia pustulosa* and *Centrostephanus longispinus,* blanched in the dark and darkened in the light and that these responses resulted from chromatophore responses. Although this observation was confirmed by Kleinholz (366), Parker was unable to substantiate such results on a closely related species, *Arbacia punctulata* (367). Parker considered it quite unlikely that a system of color change would have evolved in forms lacking inadequate photoreceptors. He considered it much more likely that color change "arose only in those forms in which the eyes and the central nervous system organ had reached such a degree of specialization as to enable its possessors to respond to the details of a luminous environment." Dambach (368), however, has recently quite clearly demonstrated that the chromatophores of *Centrostephanus longispinus* are indeed directly responsive to light stimulation, as are those of *Diadema*. Indeed, both locomotion and spine responses in *Diadema* vary with the degree of dispersion of melanin in the chromatophores (369). The melanophores lie just above the underlying photosensitive nerves that regulate spine and body movements. Thus pigmentary responses may play an important role in regulating central or other nervous system activity as they now do in the invertebrate and vertebrate retina. It is possible that this was their early evolutionary role before they became secondarily under nervous system or endocrine control. Such an interpretation of chromatophore function obviously had not been considered when Parker concluded that "chromatophores with their associated color changes are not to be expected in the simpler animals" (367).

Cnidarians

The dispersion of pigment within chromatophores of a siphonophore, *Nanomia cara,* has now been described (370). As in annelids and echinoderms, pigment is dispersed during the day and is concentrated within the chromatophores at night. How these pigment cells are controlled was not determined.

mechanisms
of hormone action

STRUCTURAL
REQUIREMENTS FOR MSH ACTIVITY

The melanophore-stimulating hormone (MSH, intermedin) is the most potent melanosome-dispersing agent known, but the entire molecule is not necessary for activity (371, 372). Some darkening of isolated frog skins *in vitro* can be induced by the centrally located pentapeptide sequence: His·Phe·Arg·Try·Gly of the MSH molecule. The structural requirements of MSH peptides for iridophore responses (reflecting platelet aggregation) parallel those needed to induce melanophore responses (373). These results imply that common features of the MSH molecule are involved in the stimulation of both iridophores and melanophores.

It is a common belief that the L-configuration of amino acids is a basic structural requirement for maximal biological activity. Some MSH peptides that contain either D-phenylalanine or D-arginine in the pentapeptide sequence have been synthesized. The D-Phe peptide is more active than the corresponding L-Phe peptide (374). Among nine stereoisomers of MSH pentapeptides, six isomers were found (375) to be as active or more active, *in vitro,* in darkening frog skins than was the all-L pentapeptide. Even more interesting was the observation that three of the isomers that were either totally inactive or nearly so reversed or inhibited the darkening action of α-MSH on frog skins. These results suggest that these "inactive" pentapeptides may be acting as antagonists of the MSH receptor. These same isomers failed to prevent darkening of frog skins in response to caffeine, indicating that the mechanisms by which MSH and caffeine disperse melanosomes differ to some extent.

THE FIRST MESSENGER-
SECOND MESSENGER
HYPOTHESIS OF HORMONE ACTION

According to the "first messenger-second messenger hypothesis" (376), hormones, such as MSH, act as first messengers and initiate their specific effects on cells by stimulating an intracellular increase in a second messenger, which is, in turn, responsible for the particular response of the effector cell. For all tissues that have been studied, this intracellular second messenger has been shown (377) to be 3', 5'-cyclic adenosine monophosphate (cyclic AMP) (Fig. 8–1). In the case of MSH action, cyclic AMP production in

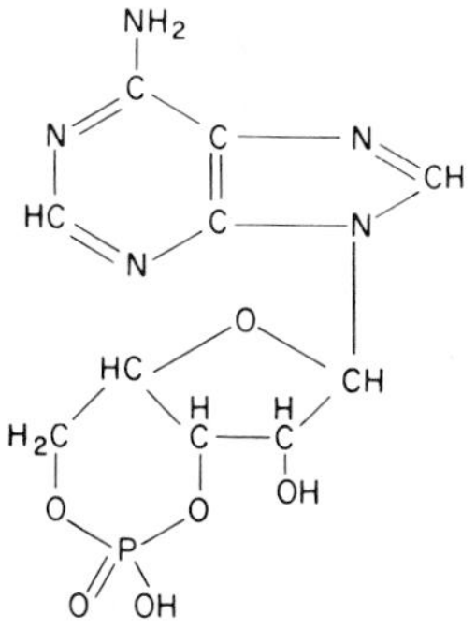

FIG. 8-1. Cyclic adenosine 3', 5'-monophosphate (cyclic AMP). The "second messenger" of hormone action.

response to this hormone would then initiate melanosome dispersion. This step is known to occur, for the addition of the second messenger, cyclic AMP, mimics the action of the first messenger, MSH, by darkening frog skins *in vitro* (21, 378, 379). The darkening response is quite slow and rather minimal, but the reason may be because of the difficulty this nucleotide might have in penetrating into cells. The dibutyryl derivative of cyclic AMP is much more effective in mimicking the action of the first messenger, and this is also true on frog skins (21, 380), where it causes a rapid and maximal darkening. Some specificity of the darkening response of frog skins to adenine nucleotides is demonstrated by the observation that skins do not darken in the presence of either 2', 3'-cyclic AMP or the noncyclic nucleotide, 5'-AMP (21). See Table 8–1. The darkening response of frog skins to either cyclic AMP or its dibutyryl analog can be considered MSH-like in that iridophores as well as melanophores respond to these nucleotides (380).

TABLE 8-1

Comparative *in vitro* responses of *Rana pipiens* skins to adenine nucleotides	
Treatment	*% Change in reflectance*
Cyclic 3', 5'-AMP	−12 ± 1.33
Dibutyryl cyclic 3', 5'-AMP	−53 ± 1.40
Cyclic 2', 3'-AMP	− 3 ± 1.11
5'-AMP	+ 1 ± 1.28
MSH	−52 ± 2.58
Ringer	+ 2 ± 0.90

Concentration of the nucleotides, 1×10^{-2} M. Concentration of MSH, 2×10^{-9} g/ml. Values are means ± S.E. Values represent greatest change in reflectance within 120 minutes after addition of the agents to the skins.

Additional evidence strongly suggesting a role for cyclic AMP as the intracellular regulator of chromatophore responses is the demonstration (381) that MSH stimulates cyclic AMP formation in dorsal, but not ventral, frog skin (Fig. 8–2). Graded concentrations of MSH produce graded increases in cyclic AMP levels within skins, and these levels return to a base value when lightened by rinsing in Ringer solution in the absence of MSH (Fig. 8–3). MSH and ACTH bring about increased concentrations of cyclic AMP in the same relative effectiveness as they exhibit in stimulating darkening of the skins.

Although these results clearly implicate a role for cyclic AMP in darkening of frog skin in response to MSH, the data do not necessarily imply that cyclic AMP synthesis occurs only within melanophores. Intermedin, as well as exogenous cyclic AMP and dibutyryl cyclic AMP, stimulates iridophore responses in dorsal frog skin (21) ; and since these reflecting cells also contribute to color change, it is quite likely that measurements of integumental changes in cyclic AMP levels also reflect changes in intracellular levels of cyclic AMP within iridophores. Iridophores are abundant in ventral frog skin where only very few melanophores are present, but these iridophores are unresponsive to MSH, ACTH, or other agents effective in darkening dorsal frog skin. One would not expect, therefore, to observe cyclic AMP changes within ventral frog skin in response to MSH treatment.

Methylxanthines, such as caffeine and theophylline, darken frog skins by dispersing melanosomes within melanophores and by aggregating reflecting platelets within iridophores. Theophylline acts synergistically with MSH in increasing the content of cyclic AMP within dorsal frog skin (381). These

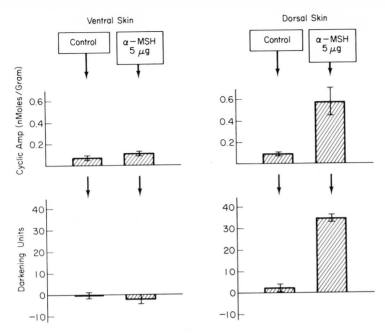

FIG. 8-2. Dorsal and ventral frog skin cyclic AMP levels in response to α-MSH *in vivo*. Either 5 μg of α-MSH in 0.5 ml 0.9% NaCl or 0.5 ml 0.9% NaCl was injected into the dorsal lymph sacs of intact frogs. Thirty minutes later cyclic AMP was measured in both the dorsal and ventral skin of the same frog. Another group of frogs were similarly injected for darkening measurement. The degree of darkening of both the dorsal and ventral skin was measured in each frog. The brackets represent the standard errors of the means. [Redrawn (381)]

FIG. 8-3. Darkening effect of α-MSH or cyclic AMP on the isolated frog skin. Mounted dorsal thigh skins were incubated in 5 ml of frog Ringer solution containing either 0.02 ng α-MSH/ml or 10 mmoles cyclic AMP. Darkening of each mounted skin was measured after 15-, 30-, and 60-minutes incubation. The brackets represent the standard errors of the means. [Redrawn from (381)]

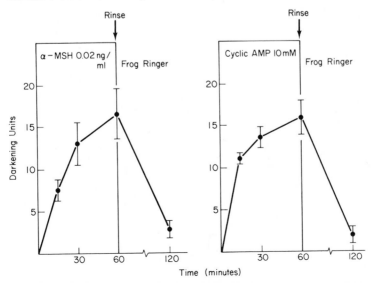

methylxanthines apparently inhibit the destruction of cyclic AMP by inhibiting a specific phosphodiesterase that normally converts cyclic AMP to
noncyclic AMP, thereby allowing an increase in endogenous cyclic AMP
levels within chromatophores. This suggestion is supported by the observation that cyclic AMP acts synergistically with theophylline in darkening frog
skins (380). Thus the mechanism by which methylxanthines darken skins
is quite different from that of MSH. Intermedin, acting as first messenger,
stimulates cyclic AMP production, whereas methylxanthines prevent the
destruction of cyclic AMP. Both agents, therefore, apparently darken skins
by producing an intracellular increase in cyclic AMP within chromatophores. Cyclic AMP has been implicated in the regulation of chromatophores
in crustaceans (309), as will be discussed later.

RECEPTOR MECHANISMS
REGULATING CHROMATOPHORE RESPONSES

There is good evidence that ACTH (382), as well as some other
peptide or protein hormones (383, 384), exerts its effect on cell membranes
rather than acting intracellularly. ACTH, for example, can be chemically
bonded to cellulose and still manifest its effects on adrenal cortical cells,
although the cellulose to which it is bound does not enter the cells (382).
ACTH contains a partial amino acid sequence similar to that of α-MSH
and is also a potent darkening agent of frog skin. Consequently, it is possible
that the receptor with which MSH interacts is a component of the plasma
membrane of melanophores and iridophores. The ability of certain peptides
of MSH to block darkening of frog skins in response to MSH, but not
darkening induced by methylxanthines (375), indicates that these peptides
may be blocking the membrane receptor or at least some step proximal to
cyclic AMP production.

Although the presence or absence of circulating MSH is the major
factor regulating pigment granule movements within vertebrate chromatophores, other hormones may actively antagonize the action of MSH on
pigment cells. In the lizard *Anolis carolinensis,* such catecholamines as
epinephrine and norepinephrine cause a rapid *in vivo* or *in vitro* lightening
of MSH-darkened skins (119). It is well recorded in the voluminous pharmacological and physiological literature that catecholamines mediate their
responses through what are referred to as adrenergic receptors. These are
the cellular receptive substances (385) that under normal physiological
stimulation would respond to catecholamines of either adrenomedullary
or sympathetic nervous origin. These receptors are generally characterized

as being of two types, *alpha* or *beta,* each of which usually controls responses that are opposite in nature to the other (386). For example, stimulation of *alpha* adrenergic receptors of vertebrate smooth muscle normally leads to contraction, whereas stimulation of beta adrenergic receptors leads to relaxation. Cells may possess either *alpha* or *beta* receptors, or both. *Alpha* receptors, when present, usually dominate over *beta* receptors. Stimulation of adrenergic receptors may lead to rapid mechanical-like contractile events of muscle or to less rapid but quite important metabolic events, such as the release of glucose from the liver or the release of free fatty acids from adipose tissue.

In *Anolis, alpha* adrenergic receptors of melanophores regulate melanin granule aggregation in response to endogenous catecholamines (387), apparently of adrenomedullary origin, for *Anolis* melanophores are not innervated (118). In nature, the release of these catecholamines results in the so-called excitement pallor when these lizards are stressed. The rapid color change from a dark chocolate brown to a bright green is the basis for the chameleonlike chromatic responses of these lizards. It will probably be shown that in all vertebrates possessing the ability to change rapidly from a dark to a light color, the melanophores of these animals possess *alpha* adrenergic receptors. In teleosts, there is good evidence that the sympathetic nervous system regulates melanosome aggregation and that the neurotransmitter mediates its effects through *alpha* receptors (388) possessed by the melanophores.

A number of pharmacological agents (such as Dibenamine, phentolamine, ergotamine) antagonize the *alpha* adrenergic receptor and prevent effector cell stimulation, thus blocking the anticipated response (Fig. 8–4). Long ago it was observed that pretreatment of tissues with ergot (a substance containing a number of alkaloids, including ergotamine) not only prevented smooth muscle contraction in response to catecholamines but was also responsible for what Dale (389) referred to as an "epinephrine reversal." This unexpected result is now explained by the fact that tissues may possess *beta* adrenergic receptors as well as *alpha* adrenergic receptors. Antagonism of the *alpha* receptors, which would normally dominate over *beta* receptors, allows for *beta* receptor expression in response to catecholamines. This point is clearly demonstrated by the response of *Anolis* melanophores after *alpha* receptor blockade. Whereas norepinephrine normally lightens *Anolis* skins, whether previously darkened by MSH or not, this same hormone causes melanosome dispersion after *alpha* receptor blockade. Because this dispersion can be blocked or reversed by propranolol or dichloroisoproterenol, potent *beta* receptor blocking agents, it is concluded that melanosome dispersion in response to catecholamines is mediated through stimulation of the *beta* receptor (Fig. 8–5). This conclusion is further sub-

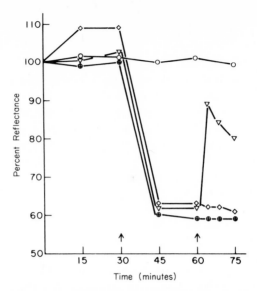

FIG. 8-4. Response of MSH-darkened *Anolis* skins to norepinephrine after *alpha* adrenergic blockade. After the initial base photometric reading, one group of skins (◇) was incubated in ergotamine (10^{-5} M) for 30 minutes, and three other groups were maintained in Ringer solution. Then MSH (2×10^{-9} g/ml) was added (↑) to the group of ergotamine-incubated skins (◇) and to two Ringer groups (▽ and ⊕). Another Ringer group (○) was maintained as a control. At 60 minutes (↑) norepinephrine (10^{-5} M) was added to two groups of skins (▽ and ◇). One group was allowed to remain as the MSH control (⊕). Results are means of the reflectance measurements from eight skins for each point on the graph. [From (387)]

FIG. 8-5. Comparative *in vitro* response of *Anolis* skins to norepinephrine (NE) and isoproterenol (ISO) after *alpha* or *beta* adrenergic blockade. After a 45-minute preincubation in dichloroisoproterenol (DCI) (2×10^{-5} M) (◆ and ○) or Dibenamine (2×10^{-5} M) (● and ▽), ISO (10^{-5} M) was added (↑) to two groups of skins (○ and ●), and NE was added (10^{-5} M) to the remaining two groups (◆ and ▽). At 90 minutes (↑) DCI (2×10^{-5} M) was added to ● and ▽. Each point on the graph is the mean of the reflectance measurements from the eight skins in the group. [From (387)]

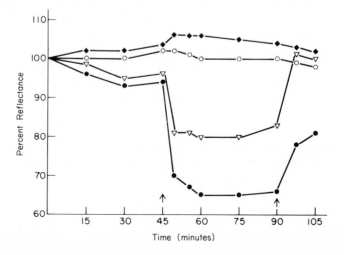

stantiated by the fact that isoproterenol, a synthetic and specific *beta* agonist, darkens *Anolis* skins after *alpha* blockade, but its melanosome-dispersing action is blocked and reversed by *beta* blockers.

Actually, Spaeth and Barbour in 1916 (255) obtained similar results in their studies on the melanophores of the isolated scale of the teleost, *Fundulus heteroclitus*. At that time, however, because adrenergic receptors had not been described, the results of their experiments could not be clearly interpreted. These early results do indicate that, as more recent experiments appear to verify, *beta* receptors of teleost melanophores also regulate melanosome dispersion.

The adaptive responses of some fish to either a light- or a dark-colored background are very rapid, being completed within a matter of minutes. *Alpha* adrenergic receptors possessed by melanophores actively mediate the lightening response that results from melanosome aggregation when fish are adapted to a light-colored background. Although there is apparently some evidence for a neuronal regulation of melanosome dispersion in response to a dark-colored background, the nature of the neurotransmitter is unknown (390). Because of the early recognition that the effects of sympathetic and parasympathetic nervous systems were usually antagonistic to each other, Parker championed the view that melanosome dispersion is mediated by acetylcholine (115). This suggestion has been incorporated into many discussions of teleost melanophore regulation. Although acetylcholine is clearly involved in vertebrate ganglionic transmission in both the parasympathetic and sympathetic nervous systems, there is no evidence that the parasympathetic system and acetylcholine play a direct role in the regulation of teleost melanophores (391). In consideration of the fact that teleost melanophores may possess both *alpha* and *beta* adrenergic receptors and that catecholamines can stimulate either of these receptors to aggregate or disperse melanosomes (391), it is interesting to speculate as to whether or not the release of a single neurotransmitter substance, norepinephrine, or other catecholamine from sympathetic nerve endings might normally regulate both aggregation and dispersion of melanosomes. This process would necessitate the preferential and temporal stimulation of either the *alpha* or *beta* adrenergic component of the melanophores under normal physiological conditions.

In *Anolis carolinensis*, there is a mosaic population of melanophores having both *alpha* and *beta* adrenergic receptors, and others apparently having only *beta* receptors (99) or more dominant *beta* adrenergic receptors. This explains why norepinephrine and epinephrine can either lighten or darken skins. For example, under conditions of stress, lizards will lighten in response to adrenal catecholamine stimulation of the *alpha* receptors possessed by the majority of the melanophores. By contrast, in lizards initially very light (green) in color, stress will induce a mottling of the skin, resulting from melanosome dispersion within localized populations (spots) of melanophores. Such stimulation of *beta* receptors provides an adequate explanation for what some workers (136) have referred to as "excitement-darkening."

Injected catecholamines may lighten dark-background adapted frogs or toads but are generally quite ineffective in lightening skins, *in vitro* (36). This situation contrasts with their dramatic *in vitro* lightening effect on *Anolis* skins. The minimal *in vitro* lightening of frog skins is, as for the lizard *Anolis carolinensis,* mediated through *alpha* adrenergic receptors possessed by the melanophores (98, 392). In addition, frog iridophores possess *alpha* receptors, and catecholamine stimulation of these receptors redisperses their reflecting platelets, thus contributing to the lightening of the skins (283, 392).

The response of the melanophores of the spadefoot toad, *Scaphiopus couchi,* differs from that of most amphibians previously studied in that MSH-darkened skins become even darker, rather than lighter, *in vitro,* in response to norepinephrine or epinephrine (380). Moreover, these catecholamines darken skins maintained in Ringer solution to about the same extent as does MSH. This darkening is regulated through *beta* adrenergic receptors, and the failure of MSH-darkened skins to lighten in response to catecholamine stimulation is due to the absence of demonstrable melanophore *alpha* adrenergic receptors. Thus species differences in chromatophore responses to catecholamine stimulation are related to the nature of the adrenergic receptors possessed by their chromatophores. Within the species *Rana pipiens,* geographical races of this frog exhibit physiological (developmental, 393) as well as morphological differences. Skins from frogs of northern United States origin (Wisconsin) respond differently to catecholamines *in vitro* than do individuals of a more southern (Mexican) origin. Skins from northern frogs lighten in response to catecholamines, whereas southern frogs darken, apparently because their chromatophores lack *alpha* adrenergic receptors (283, 392). Such darkening results from *beta* adrenergic stimulation of both melanosome dispersion and reflecting platelet aggregation, which again illustrates the parallelism of response between melanophores and iridophores pointed out earlier. These results on the toad, *Scaphiopus couchi,* and the frog, *Rana pipiens,* explain the basis for excitement-darkening in the South African Clawed Toad, *Xenopus laevis.* Most if not all, melanophores of this toad lack *alpha* adrenergic receptors but do possess *beta* adrenergic receptors (394).

Comparative studies of adrenergic regulation of chromatophore responses of a variety of other amphibians and reptiles will undoubtedly reveal a similar paradox in chromatophore regulation. Even within a single individual frog, toad, or lizard, the nature of adrenergic receptor regulation may differ between individual chromatophores. Melanophores immediately adjacent to each other within the skin may possess either *alpha* and *beta* adrenergic receptors or, apparently, only *beta* receptors. Epidermal melanophore responses to catecholamines (*X. laevis* and *R. pipiens*) are regulated solely through *beta* receptors, for *alpha* receptors are absent. These differences in adrenergic receptor characteristics of integumental chromatophores may be important in pattern regulation. It is interesting, in this context, that both the iridophores and the interspot melanophores with which they are associated in the *dermal chromatophore unit* (10) possess adrenergic receptors that, when stimulated, produce chromatic effects which complement each other. In other words, stimulation of *alpha* receptors aggregates melanosomes and disperses reflecting platelets; stimulation of *beta* receptors disperses melanosomes and aggregates reflecting platelets.

RECEPTORS AND CYCLIC AMP: GENERALIZED MODEL FOR CHROMATOPHORE CONTROL

Both MSH and *beta* adrenergic receptor stimulation disperses melanosomes and aggregates reflecting platelets. Since MSH stimulates the synthesis of cyclic AMP within these chromatophores, the question arises as to whether *beta* adrenergic stimulation also increases the intracellular level of cyclic AMP within pigment cells. Numerous studies on a variety of tissues have shown that they contain cells that are regulated both by adrenergic mechanisms and by peptides or protein hormones. Although no answer is as yet available for pigment cells, it is known that stimulation of *beta* adrenergic receptors increases cyclic AMP synthesis (395). Moreover, a great deal of evidence now suggests that in those tissues where *beta* receptors occur (liver, adipose tissue, heart) "These receptors are closely associated with (if not an integral component of) the adenyl cyclase system" (377). It is known that in cells possessing *beta* receptors the effects of catecholamines and methylxanthines are synergistic, thereby producing large increases in cyclic AMP. The darkening effects of catecholamines on either frog or reptilian (99) skins are also synergistic with methylxanthines, again suggesting that *beta* adrenergic receptor stimulation leads to an increase in cyclic AMP that is then responsible for melanosome dispersion and reflecting platelet aggregation. In those frogs or reptiles whose melanophores have *alpha* adrenergic receptors, the ability of catecholamines to antagonize the actions of MSH is apparently due to stimulation of *alpha* receptors, probably leading to a

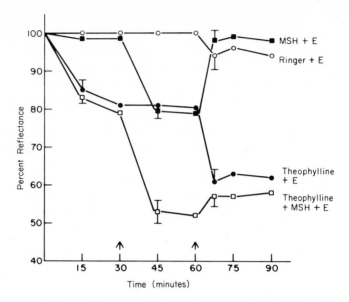

FIG. 8-6. Comparative response *in vitro* of MSH-darkened and theophylline-darkened *Anolis* skins to epinephrine. Two groups (● and □) were submaximally darkened by theophylline (10^{-3} M) for 30 minutes. Then MSH was added to a control group (■) of skins in Ringer solution and to one of the theophylline-darkened group (□) of skins. At 60 minutes, epinephrine (10^{-5} M) was added to the skins of each group and also to the skins of a control group (○) in Ringer solution. Each point on the graph is the mean of eight reflectance readings. Vertical lines indicate the standard error. [From (99)]

decrease in cyclic AMP levels. The result is a lightening of MSH-darkened skins. Stimulation of *alpha* adrenergic receptors is known to decrease the intracellular level of cyclic AMP within certain tissues (396). If, however, causes an increased darkening rather than lightening of the skins (Fig. 8–6). skins are darkened with methyxanthines, then the addition of catecholamines These results are explained by the fact that in the presence of intermedin, catecholamines probably stimulate both *alpha* and *beta* adrenergic receptors, but the effects on the *alpha* receptors dominate, thus resulting in lightening of the skins. In the presence of methylxanthines, however, the effects of catecholamines on the *beta* receptors dominate, apparently due to the inactivation of phosphodiesterase by the methylxanthines. As discussed earlier, in the absence of chromatophore *alpha* receptors, catecholamines also cause an increase in darkening of MSH-darkened skins. In this case, failure of catecholamines to lighten skins by decreasing cyclic AMP levels is directly related to the absence of chromatophore *alpha* receptors.

In searching for the molecular basis for the adrenergic receptor, it has been found that the *beta* adrenergic receptor is located within a membrane, apparently the plasma membrane (397). Stimulation of the *beta* receptor can lead to activation of adenyl cyclase, an enzyme also in the same membrane, which then converts adenosine triphosphate (ATP) to 3', 5'-cyclic adenosine monophosphate (cyclic AMP). It is not yet known whether adenyl cyclase and the *beta* adrenergic receptor are identical or are only closely related membrane components. Present evidence is that they are separate (397). It is interesting that the receptor for MSH is probably also in the membrane, but this receptor is not identical to the *beta* adrenergic receptor because *beta* adrenergic blocking agents do not block MSH-induced melanosome dispersion, except when used at very high concentrations (398). Less information is known about the nature of the *alpha* adrenergic receptor. It is not known whether *alpha* receptors are localized within the plasma membrane, and it is not understood how their stimulation leads to decreased cyclic AMP levels.

A model (377) of the second-messenger system involving adenyl cyclase is shown in Fig. 8–7. This generalized model for the control of vertebrate

FIG. 8-7. The first messenger-second messenger system for the regulation of cellular activity. [Redrawn from (377)]

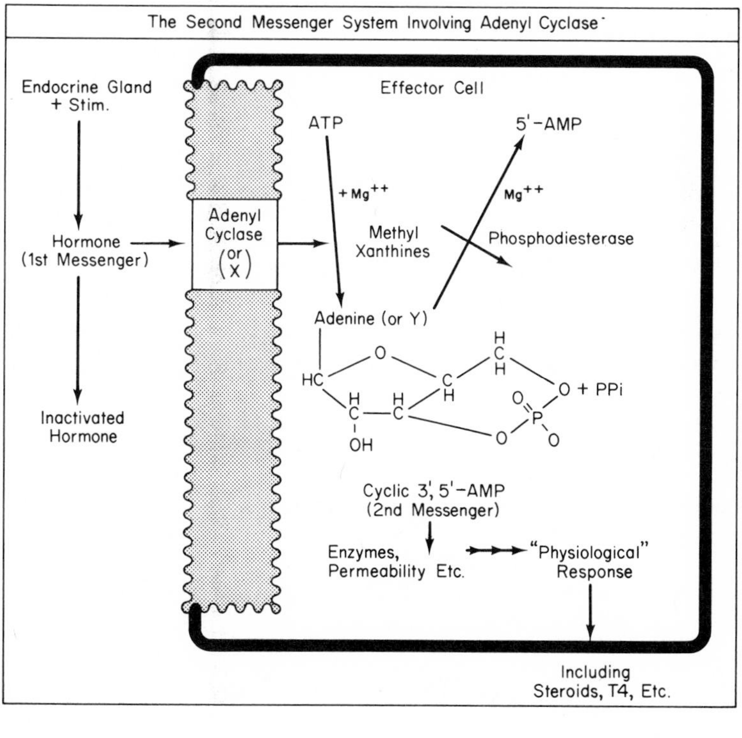

chromatophores, as well as other cellular responses, marks a great advance in our understanding of the mechanisms of pigment cell regulation.

Other hormones antagonize the actions of MSH on vertebrate chromatophores and probably do so by interacting with specific receptive substances possessed by pigment cells. For example, acetylcholine is a potent lightening agent of MSH-darkened skins of the frog, *Rana pipiens*. Preincubation of skins in atropine will block the lightening effect of acetylcholine, thus indicating that the chromatophores possess cholinergic receptors as described for other tissues. Other melatonin or serotonin receptors apparently mediate pigment granule movements within chromatophores of those species that are sensitive to these hormones. These hormones, like catecholamines, may lighten frog skins by decreasing cyclic AMP levels but apparently do so through separate, nonadrenergic receptors. The inability of such hormones as acetylcholine and melatonin to relighten frog skins darkened by methlxanthines may indicate that they normally decrease intracellular cyclic AMP levels through a mechanism involving a stimulation of cyclic AMP phosphodiesterase activity.

THEORIES ON THE MECHANISMS OF PIGMENT GRANULE MOVEMENTS

Although cyclic AMP is clearly implicated as the second messenger controlling the events within cells regulated by endocrines, less information is available as to the exact mechanisms by which cyclic AMP mediates its effects. Within liver cells, cyclic AMP apparently activates a kinase that converts dephosphophosphorylase to the active phosphorylase, which is then responsible for the conversion of glycogen to glucose. In fat cells, cyclic AMP activates one or more lipases that initiate lipolysis, leading to glycerol and free fatty acid release. Although cyclic AMP is apparently responsible for stimulating melanosome dispersion and reflecting platelet aggregation, it is not yet known how this nucleotide regulates pigment granule movements within chromatophores.

Sol-gel transformations and microtubules

The major theory accounting for the processes of pigment granule dispersion and aggregation suggests that these movements are associated with the processes of solation and gelation (399, 400). This hypothesis is given strong support by the observation that increasing hydrostatic pressures (Fig. 8–8) cause progressive melanosome dispersion within melanophores of the

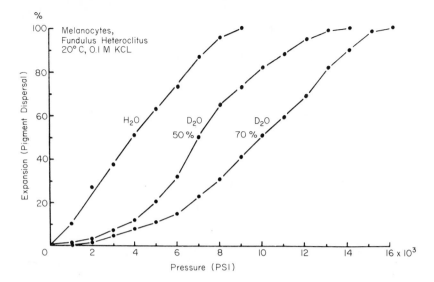

FIG. 8-8. Effects of heavy water (D_2O) on the pressure-induced expansion of melanocytes. The D_2O percentages indicate the extent to which H_2O was replaced by D_2O in the standard 0.1 M KCl test solutions. Standard deviations for the points at each specified pressure were in every case less than those shown by the vertical lines in Fig. 8-9. [Redrawn from (400)]

FIG. 8-9. Effects of temperature and D_2O on the pressure-induced expansion of melanocytes. Standard deviation values are shown by the vertical lines in the 25° C–H_2O curve. These values were maximal in the sense that the corresponding values of the other points, at each pressure level, were never greater. [Redrawn from (400)]

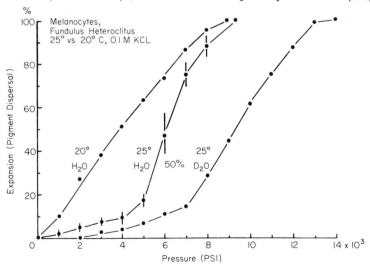

fish, *Fundulus heteroclitus,* an effect that is counteracted by increasing concentrations of heavy water (D_2O) or temperature (Fig. 8–9), factors that tend to stabilize the gelated state. This model is further supported by the observation that colchicine causes a darkening of frog skins by dispersing melanosomes (279) and aggregating reflecting platelets (225). Colchicine is thought to disrupt the oriented gelated structure of the mitotic spindle by disorganizing the orientation of the micelles in the astral rays and spindle fibers. It has been suggested that the effects of colchicine on melanophores might similarly involve a solational process resulting from a disruption of the protein units of the theorized gel structure of the pigment cells (401).

The question now arises as to whether there is evidence for ultrastructural elements, possibly related to movements of pigmentary organelles within chromatophores. In an elegant ultrastructural study of *Fundulus* melanophores (259), microtubules were observed within the dendritic processes of the pigment cells. These cytostructural elements are aligned parallel to the direction of melanosome migration and are present whether or not the melanosomes are perinuclearly aggregated or dispersed. It was suggested that these microtubular elements determine the channels in which melanosomes migrate and that pigment movements might result from some kind of interaction between the surfaces of the melanosomes and the stationary microtubules. It is interesting to speculate as to whether an interaction such as that between actin and myosin filaments of muscle may be similarly involved in melanosome movements, but there is no evidence at present to support such a hypothesis. Microtubules in this quantity or orientation have not been observed in melanophores of other teleosts (390) or in the cyclostome, *Myxine glutinosa* (125). Only small numbers of microtubules have been found in reptilian (37, 402) or amphibian dermal (10, 403) or epidermal (404) melanophores; so it is difficult at present to assign any definite pigmentary role for these cytoplasmic organelles.

Role of sulfhydryl (thiol) bonds

Sulfhydryl groups have been implicated by a number of workers as being important in the regulation of melanosome movements (405–407). Colchicine combines strongly with sulfhydryl groups and apparently causes depolymerization of the mitotic spindle protein. Substances that affect the integrity of microtubules within both melanophores and erythrophores of *Fundulus* influence pigment migration (408). These observations provide some basis for considering that the actions of colchicine on chromatophore pigment organelles may involve a similar phenomenon. Other sulfhydryl agents, such as mersalyl and *N*-ethylmaleimide (406), plus other *N*-substituted maleimides (407), have a profound darkening effect on *Anolis* skins, and, like the effects of colchicine on amphibian melanophores, their effects are irreversible. In addition to dispersing melanosomes, these maleimides also block the action of MSH on either lizard or frog melanophores. Since succinimide has no observable effects on melanophores and differs from maleimide only by the absence of the C=C bond, it would appear that this unsaturated carbon—carbon bond is responsible for the mechanism of melanosome dispersion. The failure of maleic acid, considered to be a sulfhydryl inhibitor (409), to mimic the maleimides would further suggest that the integrity of the five-membered ring in combination with the C=C bond may be important and responsible for the instability and consequent reactivity of the C=C bond (407).

The normal physiological dispersal and perinuclear aggregation of melanosomes within *A. carolinensis* melanophores in response to endocrine stimulation are very rapid events. The ability of sulfhydryl inhibitors, such as the maleimides and mersalyl, similarly to evoke rapid responses—either melanin granule dispersion or aggregation—might suggest that these agents directly affect the mechanisms by which MSH mediates its action. The irreversibility of the effects of the sulfhydryl inhibitors, which is characteristic of such agents (409), suggests that an MSH-sulfhydryl interaction in involved in melanophore regulation, as originally suggested by Horowitz for *A. carolinensis* (406).

Sulfhydryl groups have been implicated in the initial responses of a number of cell types to peptide hormones. In the action of such hormones as vasopressin (410), a disulfide bond between the hormone and its receptor may be involved, although there is evidence that the integrity of the disulfide bond of vasopressin is not necessary for its biological activity. Similarly, the sulfhydryl blocking agents mersalyl and N–ethylmaleimide block the lipolytic action of ACTH on adipose tissue; but since ACTH lacks sulfhydryl groups, it appears that the thiol requirement for its action is farther removed from the initial hormone receptor interaction. The fact that MSH also lacks thiol groups within its structure suggests that the combination of MSH with its receptor does not itself involve an hormone-sulfhydryl bond interaction.

Although mersalyl at high concentrations mimics MSH in darkening both frog (411) and lizard (406) skins, it also blocks, as well as reverses, the action of MSH when used at lower concentrations. These facts indicate that there may be a definite sulfhydryl requirement for MSH activity. Since MSH apparently acts as the first messenger to stimulate an increase of cyclic AMP, which then acts as the intracellular second messenger to mobilize pigment organelles, the sufhydryl requirement may lie between the MSH receptor and cyclic AMP synthesis. This point is clearly suggested by the observation that although mersalyl blocks MSH darkening of *Anolis* skins, it allows melanosome dispersion in response to catecholamines (398).

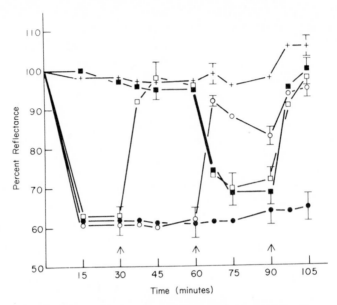

FIG. 8-10. Three groups of skins ($\square$, $\bullet$ and $\bigcirc$) were darkened by porcine *beta*-MSH (3×10^{-9} g/ml), while two groups ($\blacksquare$ and $+$) were allowed to remain as Ringer controls. At 30 minutes (arrow), mersalyl (10^{-4} M) was added to a group ($\square$) of MSH-darkened skins and to a Ringer control group ($\blacksquare$). At 60 minutes (arrow), norepinephrine (10^{-5} M) was added to a group ($\bigcirc$) of MSH-darkened skins as well as to the two groups ($\blacksquare$ and $\square$) of skins residing in mersalyl and to the Ringer control group ($+$). At 90 minutes (arrow), propranolol (10^{-4} M) was added to all the skins. Concentrations of all experimental agents are expressed as the final concentration after addition to the skins. Results are the means of the reflectance measurements from the eight skins that comprise each experimental group. Vertical lines represent standard errors of the means.

In other words, it produces an epinephrine reversal (389) by blocking an *alpha* adrenergic receptor event and allowing catecholamine stimulation of the *beta* adrenergic receptor (Fig. 8–10). This observation is important for it indicates that although both MSH and catecholamines apparently mediate melanosome dispersion by way of cyclic AMP as the second messenger, there is a specific sulfhydryl requirement for the action of MSH but not for *beta* adrenergic activity. These data provide additional evidence for the existence of separate receptors for MSH and catecholamine regulation of melanosome dispersion (398, 412). Furthermore, sulfhydryl blocking agents inhibit the lightening response of skins to catecholamines, thus implicating sulfhydryls in both melanosome aggregation and dispersion.

Since there appears to be a direct relationship between the distribution of melanosomes within epidermal melanophores and the rate of melanogenesis (40), it would seem that both these processes might, therefore, be regulated by intracellular mechanisms involving sulfhydryl compounds. Rothman and co-workers (413) demonstrated the presence of a sulfhydryl-containing compound in human epidermis that inhibited the formation of melanin from tyrosine and dopa; and they suggested that oxidation of this compound by pigmentogenic stimuli, such as ultraviolet light, by inflammatory agents, or by hormones might release the tyrosinase system from inhibition and thereby result in increased melanogenesis. There is now strong evidence to support the view that reduced glutathione is the inhibitor postulated by Rothman (414, 415). Halprin and Ohkawara (414) found that reduced glutathione is present *in vivo* in the human epidermis in concentrations shown to be inhibitory to melanin formation *in vitro*. Since hyperpigmentation following ultraviolet irradiation is preceded by a drop in glutathione reductase, these workers concluded that a similar natural mechanism might be involved in tanning. As glutathione reductase activity and reduced glutathione concentration are lower in Negroid than in Caucasian skin, it was suggested that this mechanism might account for the darker pigmentation of Negroes. Thus the enzyme, glutathione reductase, could play an important role in regulating the levels of oxidized or reduced glutathione within pigment cells (Fig. 8–11). It was also suggested that hormones or other agents might regulate melanogenesis by stimulating an increased oxidation of glutathione. It is possible, as suggested for the tyrosinase inhibitor of melanoma cells (416), that glutathione might repress the gene that regulates the synthesis of tyrosinase. Inactivation of such a repressor would lead to increased synthesis of tyrosinase followed by increased melanogenesis. Lerner et al. (417) have made the additional suggestion that thiol compounds may exert most of their inhibitory actions by combining with the copper required for enzymatic activity of the tyrosinase.

The tyrosinase by which melanin is synthesized is located within the melanosomes in the melanophore, and, therefore, the sulfhydryl groups implicated in melanogenic responses must be present within the pigment cell. As the initial effects of intermedin are probably on the plasma membrane of

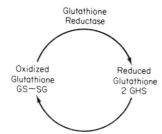

FIG. 8-11. The glutathione system. [Redrawn from (414)]

the pigment cell, a number of steps, at least one probably involving cyclic AMP, may be involved between MSH and tyrosinase activation. In the frog, *Rana pipiens,* it has been suggested that MSH-induced melanogenesis may be mediated through the release of an intracellular activator whose proteolytic action by peptide bond cleavage may convert a cytoplasmic tyrosinase to an active enzyme (418).

Ionic considerations

Since neither sulfhydryl groups nor sol-gel transformations, which, apparently may regulate pigment granule movements, are initially involved in the action of MSH, the question of what is the most proximal event in hormonal control of chromatophore activity remains to be answered. One important theory on the mechanism of MSH action concerns the ionic requirements for chromatophore regulation. Hormones might alter the permeability characteristics of chromatophore membranes to certain specific ions, which would then initiate intracellular events. Alternatively, hormones might release membrane–bound ions, which would then be responsible for stimulating enzyme activity, as is apparently the case for muscle responses.

Novales (411, 419) has reported that there is an "absolute" requirement for sodium for the action of MSH on frog melanophores. Other monovalent cations could not replace sodium and allow darkening of skins by MSH. Hypertonic "Ringer" solutions prevented the action of MSH on melanophore responses, as did hypotonic solutions. The hypotonic solutions themselves caused darkening that was only partly reversible in a few of the skins. The inhibition of MSH action in hypertonic solutions was reversible as skins slowly darkened when returned to normal Ringer solution containing MSH. The response of skins to MSH in normal Ringer after previous maintenance in hypotonic solutions was not tested. The hypothesis was formulated (411) that "MSH acts on the cell surface to bring about an increase in the Na-ion concentration of the melanophore which is proportional to the log of the concentration of MSH in the medium. Melanin dispersion is then triggered by the entry of Na ions into the cell, the amount of dispersion being proportional to the amount of Na ion entering the cell."

Although the dose-response relationship of frog skin chromatophores to MSH is affected by the sodium ion concentration of the "Ringer" solution, this fact does not necessarily suggest a direct relationship between MSH action and sodium ions. There is, however, a direct relationship between the concentration of the sodium ion, the viability of frog skins and their chromatophores, and the action of MSH. Hypotonic solutions cause frog skins to swell, skin glands to secrete, and pigment cells to burst from imbibition of water. Skins transferred to normal Ringer solution after being kept in hypotonic "Ringer" solutions are irreversibly damaged, and any remaining response to MSH is related to the number of surviving pigment cells.

Again, although "the requirement for Na is specific since neither K, choline, Li, or guanidinium was able to replace Na ion in permitting the response of frog melanophores to MSH" (419), frog skins returned to Ringer solution after immersion in potassium-"Ringer" solution do not respond to MSH. Potassium substituted for the sodium ion in the "Ringer" solution results in total disintegration of all chromatophores, as probably might be expected from the earlier observations of Spaeth (420) on teleost melanophores. Obviously one is not demonstrating a specificity for any cation requirement in these experiments. Frog skin chromatophores fail to respond to MSH in a lithium "Ringer" solution (287, 411); but after reimmersion in a normal Ringer solution, they are still unresponsive to intermedin and remain so indefinitely (421). Whether frog skins will respond to MSH in normal Ringer solution after maintenance in other sodium-substituted "Ringer" solutions has not been investigated.

Novales has found that frog skins respond to MSH in an isotonic saline solution containing only NaCl just as well as they do in Ringer solution containing the normal concentrations of calcium and potassium. In fact, there was an increased darkening response of frog skins in a calcium-free medium (422). There was a graded darkening response of frog skins to a fixed concentration of MSH in isotonic sucrose solutions containing increasing concentrations of sodium (411). This would seem to support the idea that there is indeed a requirement for the sodium ion for the action of MSH on frog chromatophores. It should be considered a real possibility, however, that sodium may play only a "permissive" role compared to that of other intracellular ions, possibly calcium or potassium, or 'both.

Intermedin-darkened *Anolis* skins pale much more rapidly· in a calcium-deficient "Ringer" solution than in a normal Ringer solution (119). Similarly, the absence of calcium from "Ringer" solution inhibits, in part, the response of *Anolis* melanophores to MSH, and the inhibition is removed (Fig. 8–12) by transferring the skins to a normal Ringer solution (423). Novales (33) has reported a calcium requirement for the action of MSH in tissue-cultured melanophores of embryonic salamanders (*Ambystoma*). Others (424, 425) have suggested that dispersion of melanosomes within melanophores of the frog, *Hyla arborea,* is regulated by an increase in the intracellular level of calcium ions and a simultaneous decrease of intracellular potassium ions. Intermedin can act on the melanophores of elasmobranchs (*Squalus* and *Mustelus*) in the absence of sodium, for lithium, choline, potassium, and magnesium all appear capable of replacing sodium in the darkening response of these dogfishes (426).

It would appear from the foregoing discussion that much remains to be done with respect to clarifying the ionic requirement for MSH action on vertebrate chromatophores. It is obvious that the theory of an obligatory sodium ion requirement for MSH action is inadequate (423). Ouabain, which is considered to block the sodium pump, neither has a direct effect on melanophores nor modifies the response of frog (421) or lizard (423) skins to MSH. Caffeine (419), theophylline (423), cyclic AMP (379), and dibutyryl cyclic AMP (423), which mimic the darkening action of MSH, have no sodium ion requirement for their actions. (See Table 8–2.) In addition, almost nothing is known relative to possible ionic requirements for active pigment granule aggregation in response to hormonal stimulation.

FIG. 8-12. *In vitro* demonstration of the immediacy of the darkening response (melanosome dispersion) of MSH-treated *Anolis* skins to added Ca^{2+}. MSH (4×10^{-9} g/ml) was added to skins immersed in Ringer solution ($\bullet + \square$) or in a Ca^{2+} free Ringer solution ($\blacksquare$). One group of skins (O) was maintained as an unstimulated Ringer control. At 30 minutes (arrow), Ca^{2+} (1 mM) was added to the Ca^{2+} free Ringer skins ($\blacksquare$) and at the same time one group of Ringer MSH-darkened skins ($\bullet$) was transferred to Ca^{2+} free Ringer (containing an identical concentration of MSH, as before). Each point on the graph is the mean of 16 measurements of reflectance. Vertical lines indicate the standard errors. [From (423)]

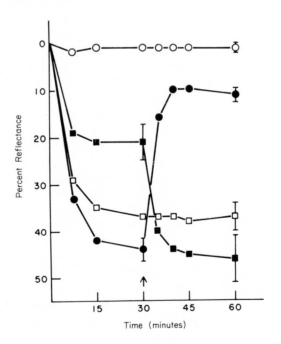

TABLE 8-2

Response (change in reflectance) of *Anolis* skins to MSH (6×10^{-9} g/ml), theophylline (10 mM), or dibutyryl cyclic AMP (DcAMP, 10 mM), in the presence or absence of calcium ion (1 mM). Each value represents the maximal mean percent reflectance change, ± standard error, of the (6) or (8) skins comprising each experimental group.

NaCl, EDTA (5 mM), MSH (6)	0 ± 0.67
NaCl, Ca^{2+}, MSH (6)	39 ± 3.69
NaCl, EDTA (5 mM), Theophylline (6)	58 ± 2.31
NaCl, Ca^{2+}, Theophylline (6)	58 ± 3.58
NaCl, EDTA (5 mM), DcAMP (8)	49 ± 2.54
NaCl, Ca^{2+}, DcAMP (8)	52 ± 1.44

Only a few preliminary observations from ionic studies on bright-colored chromatophores have been reported.

Few studies on the mechanisms of hormone action of invertebrate chromatophorotropins have been undertaken (427), apparently because little is known about the chemistry of the chromatophorotropins themselves. Some data on the ionic requirements for chromatophore responses in the fiddler crab, *Uca pugnax* (428), and the prawn, *Palaemonetes vulgaris* (429), are available.

In *Uca pugnax*, hypotonic sea water produces a reversible pigment dispersion within melanophores. The melanin-dispersing principle from the sinus gland is more effective in hypotonic solutions than in hypertonic ones. A monovalent cation is required for a maximal response of the melanophores to the melanin-dispersing principle, and sodium, potassium, and lithium are equally effective in this respect. The sodium ion permitted a much greater response of erythrophores of *Palaemonetes vulgaris* to the red pigment-concentrating hormone (RPCH) than any other cations studied. In contrast, calcium ion was most effective in permitting the effect of the red pigment-dispersing hormone (RPDH) (Fig. 8–13). Furthermore, the responses of the erythrophores to fixed concentrations of either pigment-concentrating or pigment-dispersing hormones were proportional to the respective concentrations of sodium and calcium in the extracts. Ouabain was found to inhibit the response to RPCH, whereas tetrodotoxin enhanced the response. It was found that erythrophores with maximally dispersed pigment had a transmembrane potential of 55 ± 15 mv inside negative in one set of experiments and 56 ± 4 in another set of experiments (430). No appreciable

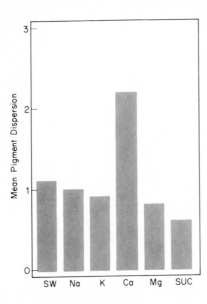

FIG. 8-13. Erythrophore responses to red pigment-dispersing hormone from extracts of the abdominal nerve cord prepared in isosmotic sea water (SW), sodium chloride (Na), potassium chloride (K), calcium chloride (Ca), magnesium chloride (Mg), and sucrose (SUC). [Redrawn from (329)]

permeability changes occur when depolarizing and hyperpolarizing currents are passed through a microelectrode within a chromatophore. Hyperpolarization of the transmembrane potential takes place in response to RPCH, and the magnitude of the hyperpolarization is directly related to the degree of pigment concentration (Fig. 8–14). Cyclic AMP dispersed the red pigment. It was suggested that the primary action of RPCH is to stimulate a pump which exchanges sodium ions from the inside of the chromatophore with potassium ions from the outside. RPDH was considered to stimulate the entry of calcium ions into the chromatophore.

The possible mechanism of action of the two red pigment-cell chromatophorotropins has been summarized (Fig. 8–15) by Fingerman (309). He explains that RPDH causes an increase in the calcium ion flux into the chromatophore and the high concentration of this ion then activates adenyl cyclase, thereby leading to the synthesis of cyclic AMP. This "second messenger" would then cause pigment granule dispersion. The red pigment-concentrating hormone requires a high external sodium concentration for its action. Ouabain is thought to block the pump that exchanges sodium ions from within the cell for potassium ions from the outside; this action would cause sodium ions to accumulate within the erythrophore because the pump cannot remove the sodium ions that enter passively. Tetrodotoxin reduces the passive influx of sodium ions into the pigment cell, thus resulting in a low internal concentration of this ion as the pump continues to remove

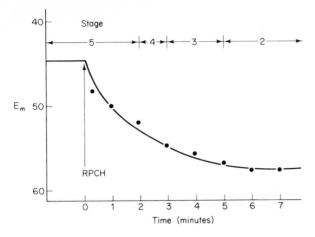

FIG. 8-14. Relationship of the transmembrane potential (Em) and chromatophore stage relative to time following application of red pigment-concentrating hormone onto the chromatophore. [Redrawn from (430)]

FIG. 8-15. Hypothetical scheme for the mechanism of action of the red pigment-dispersing and -concentrating hormones of the prawn, *Palaemonetes vulgaris.* [Redrawn from (309)]

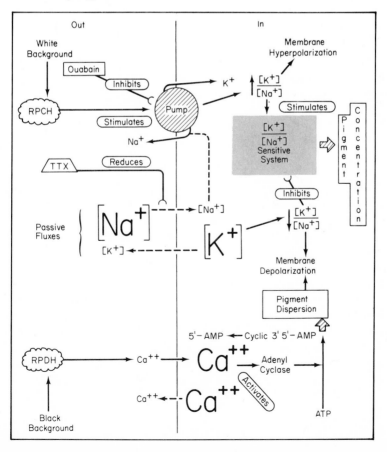

it. Since ouabain inhibits the response of the RPCH, whereas tetrodotoxin enhances it, "it is clear that for a large response to this hormone a low internal concentration of sodium ions is required." At the same time, because of the sodium dependence, it is necessary for sodium ions to be outside the erythrophore.

> The most likely explanation which ties together the observations that the transmembrane potential is predominantly potassium-dependent, the red pigment-concentrating hormone causes hyperpolarization of the transmembrane potential, and there is a sodium dependence of the red pigment-concentrating hormone, is the following: (a) the red pigment-concentrating hormone is visualized as exerting its action by stimulating the exchange pump, resulting in a high internal $K^+: Na^+$ ratio, which in turn in some manner triggers the red pigment-concentrating mechanism; (b) sodium ions enter passively while potassium ions leave passively; (c) hyperpolarization of the transmembrane potential occurs as a result of the action of the ouabain-sensitive exchange pump that produces an increase in the internal concentration of potassium ions while removing sodium ions; and (d) sodium ions are assumed to be required externally in order for the exchange pump to operate. The transmembrane potential of these erythrophores is predominantly a function of the transmembrane gradient of potassium, the concentration of potassium ions being higher inside the erythrophore than outside. The passive fluxes tend to depolarize the transmembrane potential, while the active process through the exchange pump tends to cause hyperpolarization. When the red pigment-concentrating hormone is not expressing itself, as when a high concentration of red pigment-dispersing hormone is present, the pump would operate at a low level, resulting in a low internal $K^+: Na^+$ ratio, and consequently partial depolarization of the transmembrane potential would occur. (309)

The electrophoretic theory of melanosome movement

Finally, a remaining theory for the mechanism of pigment granule movement is the electrophoretic theory of Kinosita (431). Kinosita reported that the melanosomes of the teleost (*Oryzias*) are negatively charged and that they migrate electrophoretically in the cytoplasm. This melanosome movement is in a direction opposite to that of the intracellular electrical current which results from the change in the charge of the plasma membrane that is induced by nervous stimulation. This hypothesis has not been put to a thorough test by other workers in the field; and since some data are apparently inconsistent with the theory (390), it does not have wide acceptance.

Energy requirements
in chromatophore control

A final but important consideration relative to the mechanisms of chromatophore control relates to the energy requirements necessary for pigment organelle movements. Such a consideration bears directly on the oft-discussed problem of the "active" versus the "resting" states of chromatophores. Most workers have favored the view that melanosome aggregation represents the energy-requiring state of the melanophore. Wright (279) obtained data from *in vitro* experiments on frog skin that melanophores require a source of energy from glycolysis for melanosome aggregation. Sodium iodoacetate partially inhibits the first lightening of skins darkened with MSH; but after a second darkening with MSH in the presence of iodoacetate, melanosome reaggregation is totally inhibited (Fig. 8–16). MSH was considered to disperse melanosomes by inhibiting glycolysis. Although iodoacetate interferes with glycolysis, it is possible that its effect on frog melanophores may involve a sulfhydryl inhibition other than, or in addition to, that of the glycolytic pathway. It is quite possible that an

FIG. 8-16. Response of *Rana pipiens* skins to alternate exposures to intermedin (10 units) in the presence (●) or absence (O) of iodoacetate. Intermedin present between 0–30, 60–90, 120–150, and 180–210 minutes. In the presence of iodoacetate, a partial lightening of skins takes place after the first immersion in intermedin, but subsequent blanchings (as noted for the control skins) are inhibited. [Redrawn from (279)]

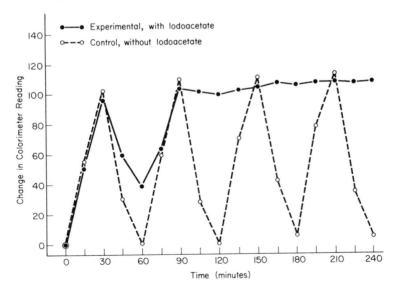

iodoacetate depletion of free thiols, such as in reduced glutathione, might be responsible for the irreversibility of iodoacetate-induced melanosome dispersion.

Only a limited number of other studies on either frog (279), lizard (119), or fish (408) chromatophores have been undertaken to determine the possible metabolic requirements for pigment organelle movements. Unfortunately, none of these studies has been done in enough depth to warrant any lengthy discussions or even any preliminary interpretations. We agree with Horowitz (119) that "the direct treatment of melanophores with ATP divulges nothing about the energy requirements of the intracellular aggregation and dispersion of melanosomes." Although there may be some

FIG. 8-17. Demonstration *in vitro* of the direct effect of temperature on responses of melanophores in *Anolis* skins. At time zero, one group (□) of skins was subjected to a warm (50° C) temperature while another group (●) was subjected to cold (4° C). At 60 minutes, MSH (3 × 10⁻⁹ g/ml) was added to the cold skins and also to a control (24° C) group (○) of skins in Ringer solution. At that same time a second group (■) of control skins in Ringer solution was subjected to a warm (50° C) temperature. At 120 minutes, norepinephrine was added to both groups of MSH-darkened skins. Each point on the graph is the mean of eight measurements of reflectance. Vertical lines indicate the standard error. [From (99)]

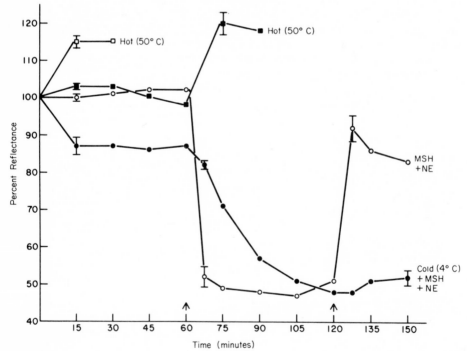

controversy as to whether or not adenosine triphosphate (ATP) darkens frog skins (119) as reported by some workers (405), both ATP as well as adenosine diphosphate (ADP) cause a rapid melanosome aggregation, resulting in lightening of *Anolis* skins (119, 432). This situation contrasts with the darkening induced by adenosine monophosphate (AMP). It is impossible at the moment to determine whether these agents directly affect melanophores, and, if so, whether these effects are intracellularly mediated or are membrane effects. Although ATP is apparently associated with contractile responses (both contraction and relaxation) of muscle, it has been found (433) that externally applied ATP activates the membrane of the isolated sartorius muscle of the frog, with resulting twitching. Horowitz (119) attributed the ability of ATP to lighten *Anolis* skins as due to its ability to bind with calcium ions. This suggestion is consistent with the observation that a lack of calcium ion in "Ringer" solution favors the aggregation of melanosomes within *Anolis* melanophores.

Information derived from temperature effects on chromatophore responses to hormonal stimulation may provide additional information relative to the energy requirements of chromatophores. Cold darkens *Anolis* skins *in vitro* (Fig. 8–17) and, in addition, decreases the rate of but does not abolish the darkening response of skins to MSH. The rapid lightening response of MSH-darkened skins to norepinephrine under normal temperature conditions is totally abolished at cold temperatures (99). These observations would seem consistent with a suggestion that melanin granule aggregation, rather than dispersion, requires a readily available source of energy.

Although a great deal of information has been obtained relative to the possible mechanisms regulating chromatophores, much still remains to be learned about the precise nature of such control. There is a great need for a clear understanding of the biochemical processes involved in pigment organelle movements to parallel the similar wealth of knowledge available from studies on other systems, such as muscle and nerve. Unfortunately, the diffuse nature of chromatophore tissue makes such a biochemical approach at present a difficult one.

CHAPTER

9

perspectives

Subsequent to the review of a large segment of scientific literature such as this, it is more than appropriate to terminate with a chapter about perspectives. At the outset of such a survey, one examines the literature of his field with a judgment flavored by his own immediate research interests. Such a vantage point is usually out of perspective. As the review progresses and more and more papers in a variety of disciplines are examined in a critical manner, one cannot help but gain a realization of the importance of many other facets of research that are only peripherally related to one's own immediate competence. In short, we develop an experience that permits viewing the subject in a more proper perspective. With this justification, the present chapter will be used to emphasize certain areas of investigation that we consider important and in need of further study. In doing so, we shall take the liberty of extending some tentative hypotheses that other investigators may choose to defend or attack.

NATURE OF CHROMATOPHORES

One of the most fertile areas for future research in pigment cell biology concerns the nature of invertebrate chromatophores. Except for the elegant study of Cloney and Florey (30) on the cephalopod chromatophore organ and a few other initial observations on other invertebrate chromatophores (17, 20), we know relatively little about the cytology of these pigment cells. More is understood about the chemistry of some invertebrate pigments; however, there have been, thus far, no attempts to associate specific types of pigments with specific chromatophore organelles. Such research holds much promise in light of what is known about such associations in vertebrate chromatophores. For example, the pigmentary organelles of the fiddler crab, *Uca pugnax*, closely resemble those of vertebrate iridophores. It would be of interest to know if they, too, contain purine pigments. To our knowledge, carotenoid-containing chromatophores, such as those of *Hippolysmata* (Frontispiece, D, E, F), have not been studied at the ultrastructural level.

PIGMENT CELL DEVELOPMENT

If little is known about the composition and structure of invertebrate chromatophores, even less is known about their origin and development. Undoubtedly, as such chromatophores are examined more and more, aspects of their developmental physiology will be revealed. Hopefully, the same can be said about vertebrate chromatophores, although it is easy to be pessimistic in view of the relatively slow accumulation of information about these problems of development. As was pointed out in Chapter 4, numerous questions about pigment cell origins remain to be resolved. The use of more sophisticated techniques will probably clarify many of the problems left unsolved by the classical transplantation procedures of earlier times.

Although it is the function of this book to point out established concepts, it is possible that a contribution can be made toward advancing our knowledge of pigmentation through the presentation of a hypothetical overall view of certain developmental aspects of pigmentation. This hypothetical scheme is prompted by the recent suggestion (19, 79, 80) that primordial pigment cells are pluripotent and have the capacity to differentiate into any of the various pigment cell types. The mechanism for the pluripotentiality would revolve around the existence in pigment cells of a primordial organelle that can form any of the definitive pigment-containing structures: melanosomes, reflecting platelets, and pterinosomes. Support for this concept comes mainly from reports describing pigment cells of one type containing subcellular organelles of another chromatophore type (19, 79, 80) and from the discovery of organelles that are composed of more than one pigment (29, 80). From this perspective, new experimental approaches may suggest themselves. First, there can be little doubt that at least some chromatophore determination takes place before neural crest cells begin migration. Determinative factors operating at this level could alter the relative numbers of each type of chromatophore that forms by influencing the primordial organelle to differentiate in one direction rather than in another. Such a concept could help explain the frequent occurrence of melanoids among amphibians (434). In comparison to normal animals, these individuals contain many more melanophores and many fewer xanthophores and iridophores. Perhaps this condition results from the conversion of prospective pterinosome and reflecting platelet organelles to melanosomes. Xanthophores adjacent to iridophores sometimes contain reflecting platelets in the part of the cell closest to the iridophores, and, similarly, melanosomes have been found in iridophores adjacent to a melanophore layer (19). It would not be too surprising if these mixed chromatophore types resulted from the exposure of the individual cells to developmental cues for both cell types; the polarity in distribution of organelles within a given cell would result from the influence from the adjacent chromatophore layer. A final important consideration is that some chromatophore components are determined but do not express themselves unless some stimulus is present. Such a stimulus could be a hormone, as in the example mentioned earlier wherein iridophores are present in tissues only during the absence of MSH.

The suggestion that developmental stimuli exist in the environment in which a chromatophore differentiates raises another important problem in the development of pigmentation. This is the question of pattern formation, especially on an animal like *R. pipiens* where there is a point-to-point correspondence between dermal and epidermal chromatophore pattern. What is the nature of the influence that ensures such conformity of pigmentary expression in dermis and epidermis alike?

PIGMENT BIOCHEMISTRY

Because of the large variation in the kinds of pigments found among invertebrates and poikilothermic vertebrates, many more biochemical studies on the pigments of these forms are necessary before we can understand their full biological significance. Of special importance is the need to know more about pigments found in cells whose physiology is at least somewhat understood. The purine and pteridine pigments of iridophores and xanthophores present a particular urgency because of the fact that a reciprocal relationship exists in their syntheses. Moreover, the synthesis of each of these pigments is dependent on MSH levels (51, 435). Hypophysioprivic larvae contain large amounts of purines and only small quantities of pteridines, whereas the opposite is true of normal larvae or of hypophysioprivic larvae that have been treated with MSH. It is supposed that the synthetic pathways of these two groups of compounds are under the control of MSH (52), but studies of this problem have been hampered by a lack of knowledge about pathways of pteridine biosynthesis in lower vertebrates.

Recent progress made in our understanding of the chemistry and physiology of phaeomelanic pigments of homiotherms offers many possibilities for experimentation. We need to know more about the distribution of these pigments among all vertebrates, particularly from the standpoint of how phaeomelanins are related to the biology of the organism. Only limited information is available about the occurrence of phaeomelanins in homeo-

therms, and these pigments have not yet been discovered among poikilotherms. Red or brown melanins are found in melanophores of some
amphibians, but they have not yet been subjected to chemical analysis. Thus
it is not known whether they are eumelanic or phaeomelanic pigments.

The biochemistry of invertebrate pigments remains a fertile field for
investigation and offers opportunities for the discovery of new and important
compounds. For example, hallochrome (originally thought to be identical to
dopachrome), a red pigment occurring in the epidermis of a polychaete,
Halla parthenopeia, has been isolated and characterized (436). It is found
to be a hydroxy-methoxy-methyl-1, 2-anthroquinone and represents the
first anthroquinone pigment to be discovered that is not substituted at positions 9 and 10. Similarly, the structure of arenicochromine, a pigment obtained from the lugworm, *Arenicola marina,* has been shown to be a
trihydroxymethoxybenzpyrenequinone (437). The origin of this polycyclic
system found in these worms is intriguing to both chemists and biologists.
Anthozoans are one of the most colorful of invertebrate groups, and sur-
prisingly little is known about the nature of their pigments. Preliminary
investigations (by L. Cariello and G. Prota) have indicated that large quantities of 3-hydroxykyurenine can be isolated from two anthozoans, *Eunicella
cavolinii* and *E. stricta.* This discovery is somewhat surprising, for 3-hydroxy-
kyurenine is a key intermediate in the synthesis of ommochrome from
tryptophan. Moreover, since ommochrome has not been found in any
invertebrate more primitive than cephalopods, the question arises as to
whether ommochrome may have a wider significance than that indicated
from the early experiments on insects. It seems probable that as we learn
more about the distribution of many of these lesser-known invertebrate pigments, information will be gained that may contribute to our knowledge of
invertebrate evolution.

CHROMATOPHORE CONTROL

Important questions at all levels about the control of chromatophores
of almost all animals remain to be answered. Although it is certain that
MSH is the primary vehicle for chromatophore control among vertebrates,
our knowledge of the hormonal regulation of invertebrate systems is less
secure. Solid research efforts have revealed the importance of chromatophorotropins in crustaceans; however, their precise chemical nature has
escaped elucidation. Hopefully, continuing work on these problems will soon
resolve not only questions about the nature of chromatophorotropins but
will also provide further clues as to how these hormones affect specific
chromatophores. Fingerman (309) has already provided a hypothetical
scheme in this regard that should help orient further investigations.

Among vertebrates, the important matter of control of MSH release commands much attention and is of continuing interest for several reasons. Considerable evidence supports the contention that the control of MSH release from the pars intermedia is based on inhibitory mechanisms. Control by adrenergic inhibition has been especially emphasized. At the same time, however, there is also reason to believe that in certain cases the control of MSH release might result from the stimulation of the pars intermedia. Various hypothalamic factors that either inhibit or stimulate MSH release have been discovered, but their roles during normal pars intermedia function are unknown. It is possible that the differences in the control of MSH release are, in part, a reflection of species variation; the answer to this question can only come from further experimentation on a diversity of vertebrates.

One of the more intriguing problems of pigmentation in vertebrates concerns the nature of the mechanism through which hormones mediate chromatophore changes. As yet these mechanisms are far from being understood. However, as a result of the extrapolation of recent advancements in pharmacology and biochemistry to pigment cell physiology, it is possible to be optimistic that a clarification of these problems will be forthcoming. One of the keys to understanding will come when more knowledge is gained about the interrelationships that seem to exist between the receptors that are indeed involved in vertebrate chromatophore regulation.

Apparently the fundamental element in chromatophore regulation is cyclic AMP, whose intracellular level is regulated through each of at least three separate receptor mechanisms: the MSH, or peptide receptor, the *alpha* adrenergic receptor, and the *beta* adrenergic receptor. It is generally considered that peptide hormones mediate their responses by reacting with a receptor on the cell surface, and this is probably true for MSH. Presumably the MSH (first messenger) response requires the intracellular buildup of cyclic AMP, the second messenger. As was pointed out earlier, MSH effects are specifically inhibited by certain sulfhydryl agents, and, interestingly enough, these agents block the *alpha* adrenergic receptor. In addition, haloalkylamine *alpha* adrenergic blocking agents (e.g., Dibenamine) also block the MSH response (398), which suggests a possible relationship between the *alpha* adrenergic receptor, sulfhydryl groups, and the MSH receptor. This point is interesting in view of the fact that stimulation of the *alpha* receptor (by catecholamines) leads to melanosome aggregation in melanophores, whereas stimulation of the MSH receptor results in melanosome dispersion. These contrasting responses can be explained by the assumption that a sequence of steps exists between the receptor and the final response and that the events initiated at the MSH receptor are different than those originating at the *alpha* receptor except for a common step that is sensitive to certain sulfhydryl blocking agents and *alpha* adrenergic blocking agents (which themselves may be sulfhydryl inhibitors). It is possible that stimulation of the *alpha* adrenergic pathway could lead to a decrease in cyclic AMP levels through an enhancement of phosphodiesterase activity, whereas activation of the MSH pathway might elevate cyclic AMP activity by inhibiting this same enzyme. Most workers have probably assumed that MSH increases cyclic AMP by a direct activation of adenyl cyclase, even though such data are lacking. However, it seems reasonable to suggest that MSH might, in fact, be acting through an action on phosphodiesterase, the enzyme that destroys cyclic AMP. If MSH were to inhibit the activity of this enzyme, this action should, theoretically, lead to an intrinsic intracellular increase in cyclic AMP (412).

Melanophore stimulation through the *beta* adrenergic receptor also leads to melanosome dispersion, presumably through the buildup of intracellular levels of cyclic AMP. Unlike either the *alpha* receptor or the MSH receptor, the *beta* receptor is not affected by sulfhydryl blocking agents, and it probably operates through the direct enhancement of adenyl cyclase activity leading to the synthesis of cyclic AMP. Left entirely unanswered at present is the possible ionic requirements for receptor activation and/or adenyl cyclase or phosphodiesterase activity.

The ever-present question of how melanosomes disperse or aggregate within the melanophore is still unresolved. Future studies need to determine the mechanism by which cyclic AMP regulates melanosome dispersion and possibly, therefore, melanogenesis. Moreover, the problem is now compounded by our increased knowlege that other kinds of pigment-containing organelles are found in other chromatophores. Do these organelles also disperse and aggregate in the same way that melanosomes migrate within the melanophore? Possibly studies at the level of the electron microscope will provide clues to the elucidation of these questions; however, it is likely that total resolution of such problems will require experimentation derived from several disciplines.

Although many aspects of pigmentation have been reviewed in this book, certain limitations have prevented a discussion of the functional significance of pigmentation. Nevertheless, it should be pointed out that this is a much-neglected area of investigation that is in need of study. The role of melanin pigmentation of the skin has been given some attention recently, especially from the standpoint of its possible function as a sun (ultraviolet light) screen or as a thermoregulatory mechanism for reducing energy expenditures (438, 439). The significance of bright colors, especially those of animals that live in near darkness, is not so well understood. Surely bright colors are of advantage in concealment, and it is known that many colors function as warnings to predators or are of nuptial significance. The selective advantages to animal pigmentation provide an intriguing subject for speculation.

CONCLUDING REMARKS

In this chapter we have posed a number of questions that will not be easily resolved; however, one can never be sure of the path of research progress. This point holds special significance for pigment cell biology because this field is at the crossroads of so many disciplines. Many investigators utilize pigment cells in their research because they are so readily visualized. In this way a wide variety of cogent questions are asked of the pigment cell and a wide variety of meaningful answers are provided. Consequently, an excellent exchange of views, serving to enhance our knowledge about this subject, is possible. It has been the aim of this book to present and to emphasize some of these views for the sake of continued progress in the physiology of pigmentation.

Addendum

Recent electron microscopic studies of both iridophores (*Anolis*, 440) and amphibian epidermal melanophores (441) have revealed the presence of numerous intracellular filaments. In melanophores, both microtubules and microfilaments have been implicated on the basis of morphological and pharmacological evidence to provide the motive force for melanosome aggregation and disperson, respectively (441, 442). In iridophores, intracellular filaments appear to be responsible for the ordered structural arrangement of reflecting platelets. Whether, in iridophores and other bright-colored pigment cells, they also control pigment organelle movements in response to hormonal or other stimuli remains to be determined.

Relative to the regulation of pars intermedia function, it has been suggested that enzymatic activity within the hypothalamus is responsible for the production of an MSH release-inhibiting factor (443). On the basis of chemical isolation (444) and structure-activity studies, this factor was said to be L-Pro-L-Leu-Gly-NH₂, the side chain of oxytocin. Other investigators, in contrast, failed to confirm these results and reported instead that tocinoic acid (445) and tocinamide (446), ring structures of oxytocin, are potent *in vitro* inhibitors of MSH release from the pituitaries of rat, hamster and some, but not all, anuran pituitaries. To further confuse the present state of understanding of pars intermedia control, it has now been reported that the initial pentapeptide sequence of oxytocin can stimulate release of MSH from the rat pituitary (447). Whether these various peptide structures truly represent physiologically important regulatory factors or are only of pharmacological significance remains to be clarified.

references

1. PARKER, G. H. *Animal colour changes and their neurohumours.* Cambridge University Press, London, 1948.

2. BAGNARA, J. T. Cytology and cytophysiology of non-melanophore pigment cells. *Intern. Rev. Cytol.* 20:173–205, 1966.

3. SNELL, R. S. Hormonal control of pigmentation in man and other mammals. *Advan. Biol. Skin* 8:447–466, 1967.

4. BYTINSKI-SALZ, H. Chromatophore studies. VII. The behaviour of *Bombina* melanophores during the epidermisation of the cornea. *Embryologia* 6:67–83, 1961.

5. FITZPATRICK, T. B., and A. S. Breathnach. Das epidermale Melanin-Einheit-System. *Derm. Wochsch.* 147:481–489, 1963.

6. HADLEY, M. E., and W. C. QUEVEDO, JR. The role of epidermal melanocytes in adaptive color changes in amphibians. *Advan. Biol. Skin* 8:337–359, 1967.

7. WITSCHI, E., and R. P. WOODS. Nuptial coloration of the bills of birds and their control by sex hormones. *Anat. Rec.* 64:85–89, 1936.

8. SZABÓ, G. The biology of the pigment cell. pp. 59–91. In *The biological basis of medicine,* Edward Bittar and Neville Bittar (Eds.), Academic Press, London and New York, 1969.

9. SZABÓ, G., A. B. GERALD, M. A. PATHAK, and T. B. FITZPATRICK. Racial differences in the fate of melanosomes in human epidermis. *Nature* 222:1081–1082, 1969.

10. BAGNARA, J. T., J. D. TAYLOR, and M. E. HADLEY. The dermal chromatophore unit. *J. Cell Biol.* 38:67–79, 1968.

11. MOYER, F. H. GENETIC effects on melanosome fine structure and ontogeny in normal and malignant cells. *Ann. New York Acad. Sci.* 100:584–606, 1963.

12. Seiji, M., K. Shimao, M. S. C. Birbeck, and T. B. Fitzpatrick. Subcellular localization of melanin biosynthesis. *Ann. New York Acad. Sci.* 100:497–533, 1963.

13. Fitzpatrick, T. B. The evolution of concepts of melanin biology. *Arch. Derm.* 96:305–323, 1967.

14. Taylor, J. D., and J. T. Bagnara. Melanosomes of the Mexican tree frog *Agalychnis dacnicolor*. *J. Ultrastruct. Res.* 29:323–333 1969.

15. Fox, H. M., and G. Vevers. *The nature of animal colours*. The Macmillan Company, New York, 1960.

16. Taylor, J. D. The effects of intermedin on the ultrastructure of amphibian iridophores. *Gen. Comp. Endocrinol.* 12:405–416, 1969.

17. Arnold, J. M. Organellogenesis of the cephalopod iridophore: cytomembranes in development. *J. Ultrastruct. Res.* 20:410–421, 1967.

18. Bagnara, J. T., M. E. Hadley, and J. D. Taylor. Regulation of bright-colored pigmentation of amphibians. *Gen. Comp. Endocrinol. Suppl.* 2:425–438, 1969.

19. Bagnara, J. T., and W. R. Ferris. Interrelationship of chromatophores. pp. 57–76. In *Biology of the normal and abnormal melanocyte*. T. Kawamura, T. B. Fitzpatrick, and M. Seiji (Eds.), University of Tokyo Press, Tokyo, 1971.

20. Green, J. P., and M. R. Neff. Fine structure of the fiddler crab epidermis. *Biol. Bull.* 137:401, 1969.

21. Bagnara, J. T., and M. E. Hadley. Control of bright-colored pigment cells of fishes and amphibians. *Amer. Zool.* 9:465–478, 1969.

22. Ortiz, E., and A. Maldonado. Pteridine accumulation in lizards of the genus *Anolis*. *Carribean J. Sci.* 6:9–13, 1966.

23. Obika, M., and J. T. Bagnara. Pteridines as pigments in amphibians. *Science* 143:485–487, 1964.

24. Obika, M. Association of pteridines with amphibian larval pigmentation and their biosynthesis in developing chromatophores. *Dev. Biol.* 6:99–112, 1963.

25. Matsumoto, J. Studies on fine structure and cytochemical properties of erythrophores in swordtail, *Xiphophorus helleri*, with special reference to their pigment granules. *J. Cell Biol.* 27:493–504, 1965.

26. Ziegler, I. Pteridine als Wirkstoffe und Pigmente. *Ergebnisse der Physiologie* 56:1–66, 1965.

27. Shoup, J. R. Development of pigment granules in the eyes of wild type and mutant *Drosophila melanogaster*. *J. Cell Biol.* 29:223–249, 1966.

28. Breathnach, A. S. The cell of Langerhans. *Intern. Rev. Cytol.* 18:1–28, 1965.

29. Bagnara, J. T., and J. D. Taylor. Differences in pigment-containing organelles between color forms of the red-backed salamander, *Plethodon cinereus*. *Z. Zellforsch.* 106:412–417, 1970.

30. Cloney, R., and E. Florey. Ultrastructure of cephalopod chromatophore organs. *Z. Zellforsch.* 89:250–280, 1968.

31. SUMNER, F. B. Vision and guanine production in fishes. *Proc. Nat. Acad. Sci.* 30:285–294, 1944.

32. LERNER, A. B., and J. D. CASE. Pigment cell regulatory factors. *J. Invest. Derm.* 32:211–221, 1959.

33. NOVALES, R. R., and B. J. NOVALES. Sodium dependence of intermedin action on melanophores in tissue culture. *Gen. Comp. Endocrinol.* 1:134–144, 1961.

34. HOGBEN, L. T., and D. SLOME. The pigmentary effector system. VI. The dual character of endocrine coördination in amphibian colour change. London, *Proc. Roy. Soc.* B 108:10–53, 1931.

35. SHIZUME, K., A. B. LERNER, and T. B. FITZPATRICK. *In vitro* bioassay for the melanocyte-stimulating hormone. *Endocrinology* 54:553–560, 1954.

36. HADLEY, M. E., and J. T. BAGNARA. Integrated nature of chromatophore responses in the *in vitro* frog skin bioassay. *Endocrinology* 84:69–82, 1969.

37. TAYLOR, J. D., and M. E. HADLEY. Chromatophores and color change in the lizard, *Anolis carolinensis*. *Z. Zellforsch.* 104:282–294, 1970.

38. GELDERN, C. E., VON. Color changes and structures of the skin of *Anolis carolinensis*. *Proc. Calif. Acad. Sci.* 10:77–117, 1921.

39. RALPH, C. L. The control of color in birds. *Amer. Zool.* 9:521–530, 1969.

40. HADLEY, M. E., and W. C. QUEVEDO, JR. Vertebrate epidermal melanin unit. *Nature* 209:1334–1335, 1966.

41. RAPER, H. S. The aerobic oxidases. *Physiol. Rev.* 8:245–282, 1928.

42. MASON, H. S. The structure of melanin. *Advan. Biol. Skin* 8:293–312, 1967.

43. NICOLAUS, R. A., and M. PIATELLI. Progress in the chemistry of natural black pigments. *Rend. Acc. Sci. Fis. Mat., Napoli* 32:83, 1965.

44. PROTA, G., and R. A. NICOLAUS. On the biogenesis of phaeomelanins. *Advan. Biol. Skin* 8:323–328, 1967.

45. PROTA, G., Structure and biogenesis of phaeomelanins. pp. 615–630. In *Pigmentation: its genesis and control*, V. Riley (Ed.), Appleton-Century-Crofts, New York, 1972.

46. BUTENANDT, A., U. SCHIEDT, E. BIERKERT. Über Ommochrome III Mitt. Synthese des Xanthommatins. *Justis Liebigs Ann. Chem.* 311:79–83, 1954.

47. FORREST, H. S. The ommochromes. pp. 619–628. In *Pigment cell biology*, M. Gordon (Ed.), Academic Press, New York, 1959.

48. KAUFMAN, S. Studies on the mechanism of enzymatic conversion of phenylalanine to tyrosine. *J. Biol. Chem.* 234:2677–2682, 1959.

49. HUTZENLAUB, W., G. B. BARLIN, and W. PFLEIDERER. A new pteridine-purine transformation. *Angew. Chem.* 8:608, 1969.

50. WEYGAND, F., H. SIMON, G. DAHMS, M. WALDSCHMIDT, H. J. SCHLIEP, and H. WACKER. Über die Biogenase des Leucopterins. *Angew. Chem.* 73:402–407, 1961.

51. BAGNARA, J. T. Chromatotropic hormone, pteridines, and amphibian pigmentation. *Gen. Comp. Endocrinol.* 1:124–133, 1961.

52. STACKHOUSE, H. L. Some aspects of pteridine biosynthesis in amphibians. *Comp. Biochem. Physiol.* 17:219–235, 1966.

53. Bagnara, J. T., and H. L. Stackhouse. Purine components of guanophores in amphibians. *Anat. Rec.* 139:292, 1961.

54. Richards, C. M., and J. T. Bagnara. Expression of specific pteridines in neural crest transplants between *Pleurodeles* and axolotl. *Dev. Biol.* 15:334–347, 1967.

55. Bagnara, J. T., and M. Obika. Comparative aspects of integumental pteridine distribution among amphibians. *Comp. Biochem. Physiol.* 15:33–49, 1965.

56. Ziegler, I. Pteridine als Winkstoffe und Pigmente. *Ergebnisse der Physiologie* 56:1–66, 1965.

57. McNutt, W. S. The incorporation of the pyrimidine ring of adenine into the isoalloxazine ring of riboflavin. *J. Biol. Chem.* 219:365–373, 1956.

58. Fox, D. L. *Animal biochromes and structural colours.* Cambridge University Press, London, 1953.

59. Gilchrist, B. M. Distribution and relative abundance of carotenoid pigments in *Anostraca* (Crustacea: Branchiopoda). *Comp. Biochem. Physiol.* 24:123–147, 1968.

60. Brush, A., and H. Seifried. Pigmentation and feather structure in genetic variants of the Gouldian finch, *Peophila gouldiae. Auk* 85:416–430, 1968.

61. Herring, P. J. The carotenoid pigments of *Daphnia magna* Straus. II. Aspects of pigmentary metabolism. *Comp. Biochem. Physiol.* 29:205-221, 1968.

62. Hata, M., and M. Hata. Carotenoid metabolism in *Artemia salina* L. *Comp. Biochem. Physiol.* 29:985–994, 1969.

63. Takeuchi, K. Specificity of carotenoid transfer in the larval medaka, *Oryzias latipes. J. Cell Physiol.* 72:43–48, 1968.

64. DuShane, G. P. An experimental study of the origin of pigment cells in Amphibia. *J. Exp. Zool.* 72:1–31, 1935.

65. Hörstadius, S. *The neural crest.* Oxford University Press, London, 1950.

66. Niu, M. C. The axial organization of the neural crest, studied with particular reference to its pigmentary component. *J. Exp. Zool.* 105:79–113, 1947.

67. Wilde, C. E., Jr. The differentiation of vertebrate pigment cells. pp. 267–300. In *Advances in Morphogenesis,* Vol. 1 M., Abercrombie and J. Brachet (Eds.) Academic Press, New York, 1961.

68. Rawles, M. E. Origin of melanophores and their role in development of color patterns in vertebrates. *Physiol. Rev.* 28:383–408, 1948.

69. Lehman, H. E., and L. M. Youngs. Extrinsic and intrinsic factors influencing amphibian pigment pattern formation, pp. 1–48. In *Pigment cell biology,* M. Gordon (Ed.), Academic Press, New York, 1959.

70. Volpe, E. P. Interplay of mutant and wild-type pigment cells in chimeric leopard frogs. *Dev. Biol.* 8:205–221, 1963.

71. Dalton, H. C. Developmental analysis of genetic differences in pigmentation in the axolotl. *Proc. Nat. Acad. Sci.* 35:277–283, 1949.

72. Mayer, T. C. Pigment cell migration in piebald mice. *Dev. Biol.* 15:521–535, 1967.

73. Mayer, T. C., and M. C. Green. An experimental analysis of the pigment defect caused by mutations at the *W* and *Sl* loci in mice. *Dev. Biol.* 18:62–75, 1968.

74. Volpe, E. P. Fate of neural crest homotransplants in pattern mutants of the leopard frog. *J. Exp. Zool.* 157:179–196, 1964.

75. Volpe, E. P. Interplay of mutant and wild-type pigment cells in chimeric leopard frogs. *Dev. Biol.* 8:205–221.

76. Chavin, W. Pituitary-adrenal control of melanization in xanthic goldfish. *Carassius auratus* L. *J. Exp. Zool.* 133:1–46, 1956.

77. Loud, A. V., and Y. Mishima. The induction of melanization in goldfish scales with ACTH, *in vitro*. *J. Cell Biol.* 18:181–194, 1963.

78. Matsumoto, J., and M. Obika. Morphological and biochemical characterization of goldfish erythrophores and their pterinosomes. *J. Cell Biol.* 39:233–250, 1968.

79. Alexander, N. J. Differentiation of the melanophore, iridophore, and xanthophore from a common stem cell. *J. Invest. Derm.* 54:82, 1970.

80. Bagnara, J. T. Interrelationships of melanophores, iridophores, and xanthophores. pp. 171–180. In *Pigmentation: its genesis and control*, V. Riley (Ed.), Appleton-Century-Crofts, New York, 1972.

81. Arnott, H. J., A. C. G. Best, and J. A. C. Nicol. Occurrence of melanosomes and of crystal sacs within the same cell of the tapetum lucidum of the stingaree. *J. Cell Biol.* 46:426–427, 1970.

82. Humphrey, R. R., and J. T. Bagnara. A color variant in the Mexican axolotl. *J. Hered.* 58:251–256, 1967.

83. Searle, A. G. *Comparative genetics of coat colour in mammals*. Academic Press, New York, 1968.

84. Mintz, B. Gene control of mammalian pigmentary differentiation. I. Clonal origin of melanocytes. *Proc. Nat. Acad. Sci.* 58:344–351, 1967.

85. Smith, P. E. The pigmentary growth and endocrine disturbances induced in the anuran tadpole by the early ablation of the pars buccalis of the hypophysis. *Am. Anat. Mem.* No. 11, 1920.

86. Bagnara, J. T. Hypophysectomy and the tail-darkening reaction in *Xenopus*. *Proc. Soc. Exp. Biol. Med.* 94:572–575, 1957.

87. Pehlemann, F. W. Der morphologische Farbwechsel von *Xenopus laevis*-larven. *Z. Zellforsch.* 78:484–510, 1967.

88. Bagnara, J. T. Analyse des transformations des ptéridines de la peau au cours de la vie larvaire et à la métamorphose chez le triton *Pleurodeles waltlii* Michah. Changements induits par l'action localisée d'implants de thyroxine-cholestérol. *C.R. Acad. Sci.* 258:5969–5971, 1964.

89. Smith, P. E. Experimental ablation of the hypophysis in the frog embryo. *Science* 44:280–282, 1916.

90. Allen, B. M. Extirpation of the hypophysis and thyroid glands of *Rana pipiens*. *Science* 44:755–757, 1916.

91. Hogben, L. T., and F. R. Winton. The pigmentary effector system. III. Colour response in the hypophysectomized frog. London, *Proc. Roy. Soc.* B, 95:15–31, 1924.

92. SWINGLE, W. W. The relation of the pars intermedia of the hypophysis to pigmentation changes in anuran larvae. *J. Exp. Zool.* 34:119–141, 1921.

93. LI, C. H. β-Lipotropin a new pituitary hormone, pp. 93–101. In *La spécificité zoologique des hormones hypophysaires et de leurs activitiés.* Centre National de la Recherche Scientifique, Paris, France, 1969.

94. GESCHWIND, I. I. The main lines of evolution of the pituitary hormones, pp. 385–400. In *La spécificité zoologique des hormones hypophysaires et de leurs activitiés.* Centre National de la Recherche Scientifique, Paris, France, 1969.

95. LOWRY, P. J., and A CHADWICK. Interrelations of some pituitary hormones. *Nature* 226:219–222, 1970.

96. LOWRY, P. J., and A. CHADWICK. Purification and amino acid sequence of melanocyte-stimulating hormone from the dogfish *Squalus acanthias. Biochem. J.* 118:713–718, 1970.

97. SHAPIRO, M., D. N. ORTH, K. ABE, W. E. NICHOLSON, D. P. ISLAND, and G. W. LIDDLE. Evidence for the presence of MSH other than MSH and β-MSH in tumors of patients with the "ectopic ACTH-MSH syndrome." *Clin. Res.* 18:35, 1970.

98. LERNER, A. B. Mechanism of hormone action. *Nature* 184:674–677, 1959.

99. HADLEY, M. E., and J. M. GOLDMAN. Physiological color changes in reptiles. *Amer. Zool.* 9:489–504, 1969.

100. GESCHWIND, I. I., and R. A. HUSEBY. Melanocyte-stimulating activity in a transplantable mouse pituitary tumor. *Endocrinology* 79:97–105, 1966.

101. ATWELL. W. J. On the nature of the pigmentation changes following hypophysectomy in the frog larva. *Science* 49:48–50, 1919.

102. ATWELL, W. J. Further observations on the pigment changes following removal of the epithelial hypophysis and the pineal gland in the frog tadpole. *Endocrinology* 5:221–232, 1921.

103. EAKIN, R. M., and F. E. BUSH. Development of the amphibian pituitary with special reference to the neural lobe. *Anat. Rec.* 129:279–296, 1957.

104. McGUINNESS, B. W. Melanocyte-stimulating hormone: a clinical and laboratory study. *Ann. N. Y. Acad. Sci.* 100:640–657, 1963.

105. LERNER, A. B., and J. S. McGUIRE. Melanocyte-stimulating hormone and adrenocorticotrophic hormone. *New England J. Med.* 270:539–546, 1964.

106. BROOKS, V. E. H., and R. RICHARDS. Depigmentation in Cushing's syndrome. *Arch. Intern. Med.* 117:677–680, 1966.

107. BRONSON, F. H., and S. H. CLARKE. Adrenalectomy and coat color in deer mice. *Science* 154:1349–1350, 1966.

108. GESCHWIND, I. I. Change in hair color in mice induced by injection of *alpha*-MSH. *Endocrinology* 79:1165–1167, 1966.

109. CLIVE, D., and R. S. SNELL. Effect of the *alpha* melanocyte-stimulating hormone on mammalian hair color. *J. Invest. Derm.* 49:314–321, 1967.

110. RUST, C. C. Hormonal control of pelage cycles in the short-tailed weasel (*Mustela erminea bangsi*). *Gen. Comp. Endocrinol.* 5:222–231, 1965.

111. LERNER, A. B., and J. S. McGUIRE. Effect of *alpha*- and *beta*-melanocyte-stimulating hormones on the skin colour of man. *Nature* 189:176–179, 1961.

, A. Enemar, and B. Falck. Monoamines in the hypothalamo-ystem of the mouse with special reference to the ontogenetic *ellforsch.* 89:590–607, 1968.

C. Electron-microscopic study of the pars intermedia of the the toad, *Bufo arenarum. Gen. Comp. Endocrinol.* 4:492–502,

. C. Monoamines and control of pars intermedia in the toad *Gen. Comp. Endocrinol.* 6:19–25, 1966.

F. C. Monoamines in the neurointermediate lobe of the pituitary rgentinian eel. *Naturwissenschaften* 21:565, 1967.

F. C. The secretion of intermedin in autotransplants of pars in-growing in the anterior chamber of intact and sympathectomized the toad. *Neuroendocrinology* 2:11–18, 1967.

, F. C. Further evidences for the blocking effect of catecholamines on cretion of melanocyte-stimulating hormone in toads. *Gen. Comp. rinol.* 12:417–426, 1969.

, S. K. Fine structure of the autotransplanted pituitary of the red eft, *phthalmus viridescens. Gen. Comp. Endocrinol.* 12:12–32, 1969.

T. Experimental studies on the hypothalamic control of the pars in-edia activity of the frog, *Rana nigromaculata. Neuroendocrinology* 3:25–1968.

tin, A. J., and G. T. Ross. Melanocyte-stimulating hormone activity in uitaries of frogs with hypothalamic lesions. *Endocrinology* 77:45–48, 1965.

hlemann, F. W. Experimentelle Beeinflussung der Melanophorenverteilung on *Xenopus laevis*-larven. *Zool. Anz. Suppl.* 29:571–580, 1967.

aland, L. C. Ultrastructure of the pars intermedia in relation to hypothala-mic control of hormone release. *Nueroendocrinology* 3:72–88, 1968.

Pehlemann, F. W. Ultrastructure and innervation of the pars intermedia of the pituitary of *Xenopus laevis. Gen. Comp. Endocrinol.* 9:481, 1967.

Cohen, A. G. Observations on the pars intermedia of *Xenopus laevis. Nature* 215:55–56, 1967.

. Burgers, A. C. J., K. Imai, and G. J. van Oordt. The amount of melano-phore-stimulating hormone in single pituitary glands of *Xenopus laevis* kept under various conditions. *Gen. Comp. Endocrinol.* 3:53–57, 1963.

78. Ortman, R. Cytochemical study of the physiological activities in the pars intermedia of *Rana pipiens. Anat. Rec.* 119:1–9, 1954.

179. Rust, C. C., and R. K. Meyer. Effect of pituitary autographs on hair color in the short-tailed weasel *Gen. Comp. Endocrinol.* 11:548–551, 1968.

180. Howe, A., and A. J. Thody. The effect of hypothalamic lesions on the melanocyte-stimulating hormone content and histology of the pars intermedia of the rat pituitary gland. *J. Physiol.* 201:25P–26P, 1969.

181. Bal, H., and P. G. Smelik. Effect of hypothalamic lesions of MSH content of the intermediate lobe of the pituitary gland in the rat. *Experientia* 23:759–760, 1967.

112. Chavin, W. Pituitary hormones in melanogenesis, pp. 63–83. In *Pigment cell biology,* M. Gordon (Ed.). Academic Press, New York, 1959.

113. Pickford, G. E. Melanogenesis in *Fundulus heteroclitus. Anat. Rec.* 125:603–604, 1956.

114. Pickford, G. E., and B. Kosto. Hormonal induction of melanogenesis in hypophysectomized killifish (*Fundulus heteroclitus*). *Endocrinology* 61:177–196, 1957.

115. Parker, G. H. Chemical control of nervous activity. C. Neurohormones in lower vertebrates, pp. 633–656. In *The hormones, Vol. II,* G. Pincus and K. V. Thimann (Eds.), Academic Press, New York, 1950.

116. Kleinholz, L. H. The melanophore-dispersing principle in the hypophysis of *Fundulus heteroclitus. Biol. Bull.* 69:379–390, 1935.

117. Kleinholz, L. H. Studies in reptilian colour changes. II. The pituitary and adrenal glands in the regulation of the melanophores of *Anolis carolinensis. J. Exp. Biol.* 15:474–491, 1938.

118. Kleinholz, L. H. Studies in reptilian colour changes. III. Control of the light phase and behaviour of isolated skin. *J. Exp. Biol.* 15:492–499, 1938.

119. Horowitz, S. B. The energy requirements of melanin granule aggregation and dispersion in the melanophores of *Anolis carolinensis. J. Cell. Comp. Physiol.* 51:341–357, 1958.

120. Noble, G. K., and H. T. Bradley. The relation of the thyroid and the hypophysis to the moulting process in the lizard, *Hemidactylus brookii. Biol. Bull.* 64:289–298, 1933.

121. Rahn, H. The pituitary regulation of melanophores in the rattlesnake. *Biol. Bull.* 80:228–237, 1941.

122. Canella, M. F. Note di fisiologia dei cromatofori dei vertebrati peciilotermi, particolarmente dei lacertili. Monitore Zool. Itali. 71:430–480, 1963.

123. Young, J. Z. The photoreceptors of lampreys. II. The functions of the pineal complex. *J. Exp. Biol.* 12:254–270, 1935.

124. Eddy, J. M. P., and R. Strahan. The role of the pineal complex in the pig-mentary effector system of the lampreys, *Mordacia mordax* (Richardson) and *Geotria australis* Gray. *Gen. Comp. Endocrinol.* 11:528–534, 1968.

125. Holmberg, K. Ultrastructure and response to background illumination of the melanophores of the Atlantic hagfish, *Myxine glutinosa* L. *Gen. Comp. Endocrinol.* 10:421–428, 1968.

126. Larsen, L. O. Effects of hypophysectomy in the cyclostome, *Lampetra fluviatilis* (L) Gray. *Gen. Comp. Endocrinol.* 5:16–30, 1965.

127. Parker, G. H., and H. Porter. The control of the dermal melanophores in elasmobranch fishes. *Biol. Bull.* 66:30–37, 1934.

128. Gorbman, A., and H. A. Bern. *A textbook of comparative endocrinology.* John Wiley and Sons, New York, 1962.

129. Kleinholz, L. H., and H. Rahn. The distribution of intermedin: a new biological method of assay and results under normal and experimental condi-tions. *Anat. Rec.* 76:157–172, 1940.

130. Witschi, E. Vertebrate gonadotrophins. *Mem. Soc. Endocrinol.* 4:149–165, 1955.

131. Ralph, C. L., D. L. Grinwich, and P. F. Hall. Studies of the melanogenic response of regenerating feathers in the weaver bird: comparison of two species in response to two gonadotrophins. *J. Exp. Zool.* 166:283–288, 1967.

132. Groenendijk-Huijers, M. M. Development of black down feathering in hybrid chick embryos (Cross New Hampshire ♂ × Light Sussex ♀) after pituitary implantation. *Experientia* 24:501–503, 1968.

133. DuShane, G. P. Neural-fold derivatives in the Amphibia. Pigment cells, spinal ganglia, and Rohon-Beard cells. *J. Exp. Zool.* 78:485–503, 1938.

134. Stoppani, A. O. M., P. F. Pieroni, and A. J. Murray. The role of peripheral nervous system in color changes of *Bufo arenarum* Hensel. *J. Exp. Zool.* 31:631–638, 1954.

135. Bagnara, J. T. Hypophyseal control of guanophores in anuran larvae. *J. Exp. Zool.* 137:265–284, 1958.

136. Sand, A. The comparative physiology of colour response in reptiles and fishes. *Biol. Rev.* 10:361–382, 1935.

137. Ford, E. A contribution to our knowledge of the life-histories of the dogfishes landed at Plymouth. *J. Mar. Biol. Assoc.* 12:468–505, 1921.

138. Hogben, L. T. The pigmentary effector system. VII. The chromatic function in elasmobranch fishes. London, *Proc. Roy. Soc.* B, 120:142–158, 1936.

139. Vilter, V. Recherches histologiques et physiologiques sur la fonction pigmentaire des sélaciens. *Bull. Soc. Sci. Arcachon.* 34:65–136, 1937.

140. Zondek, B., and H. Krohn. Hormon des Zwischenlappens der Hypophyse (Intermedin). I. Die Rotfarbung der Elritze als Testobjekt. *Klin. Wochsch.* 2:405–408, 1932.

141. Zondek, B., and H. Krohn. Hormon des Zwischenlappens der Hypophyse (Intermedin). II. Intermedin im Organismus (Hypophyse, Gehirn). *Klin. Wochsch.* 2:849–853, 1932.

142. Zondek, B., and H. Krohn. Hormon des Zwischenlappens der Hypophyse (Intermedin). III. Zur Chemie, Darstellung und Biologie Intermedins. *Klin. Wochsch.* 2:1293–1298, 1932.

143. Zondek, B., and H. Krohn. Ein Hormon der Hypophyse. Zwischenlappenhormon (Intermedin). *Naturwissenschaften* 20:134–136, 1932.

144. Sage, M. Control of prolactin release and its role in color change in the teleost *Gillichthys mirabilis*. *J. Exp. Zool.* 173:121–128, 1970.

145. Fries, E. F. B. Iridescent white reflecting chromatophores (antaugophores, iridoleucophores) in certain teleost fishes, particularly in *Bathygobius*. *J. Morphol.* 103:203–254, 1958.

146. Waring, H. *Color change mechanisms of cold-blooded vertebrates.* Academic Press, New York, 1963.

147. Mitchell, G. M., K. Dierst-Davies, and F. W. Landgrebe. *In vivo* antagonism of melanophore-dispersing hormone by melatonin in adult *Xenopus laevis*. *Proc. Roy. Soc. Med.* 62:891–892, 1969.

148. Bradshaw, S. D ... activity of melan... *zoologique des hor...* de la Recherche S...

149. Lerner, A. B., J. D. ... tion of melatonin; the ... *Chem. Soc.* 80:2587, ...

150. Stoppani, A. O. M. N... *arenarum* Hensel. *Endocr...*

151. Hadley, M. E. Cytophysic... *pipiens.* Ph.D. Thesis, Brow... Mich, 1966.

152. Etkin, W. On the control of ... the pituitary by the hypothalam... 1941.

153. Mazzi, V. Sulla presenza e sul ... ipothalamo-ipofisaire nel lobo int... *Monitore Zool. Ital.* 62:1–8, 1954.

154. Rowlands, A. The influence of wate... pigmentary system of the common t... 31:151–160, 1954.

155. Guardabassi, A. The hypophysis of *X...* removal of the anterior hypothalomus. *C...* 1961.

156. Etkin, W. Hypothalamic inhibition of the ... frog. *Gen. Comp. Endocrinol. Suppl.* 1:70–...

157. Etkin, W. Neurosecretory control of the p... *Endocrinol.* 2:161–169, 1962.

158. Dierst, K. E., and C. L. Ralph. Effect of h... melanophores in the frog. *Gen. Comp. Endocrino...*

159. Jorgensen, C. B., and L. O. Larsen. Nature of t... pars intermedia function in amphibians: rate of ... denervation. *Gen. Comp. Endocrinol.* 3:468–472, 196...

160. Dierst-Davies, K., C. L. Ralph, and J. L. Pechersk... cological agents on the hypothalamus of *Rana pipien...* control of skin melanophores. *Gen. Comp. Endocrinol.* ...

161. Sawyer, C. H. Cholinergic stimulation of the release of mel... by the hypophysis in salamander larvae. *J. Exp. Zool.* 106:...

162. Enemar, A., and B. Falck. On the presence of adrenergic ne... intermedia of the frog, *Rana temporaria. Gen. Comp. End...* 583, 1965.

163. Hillarp, N. A., and D. Jacobsohn. Über die Innervation der Ade... und ihre Beziehungen zur gonadotropen Hypophysenfunktion. *Kun... Sällsk. Handl. N.F.* 54:1–25, 1943.

164. Bjorklund, A... hypophyseal ... aspects. Z. Z...

165. Iturriza, F... pituitary of ... 1964.

166. Iturriza, ... pituitary. ...

167. Iturriza, ... of the A...

168. Iturriza... termedi... eyes of ...

169. Iturriza... the s... Endo...

170. Masu... Noto...

171. Ito, ... ter... 33,...

172. Ka... p...

173. P...

174. S...

175.

176.

17...

182. KNOWLES, F. 1965. Evidence for a dual control, by neurosecretion, of hormone synthesis and hormone release in the pituitary of the dogfish, *Scylliorhinus stellaris*. London, *Phil. Trans. Roy. Soc.* B, 249:435–456.

183. MEURLING, P., and A. BJÖRKLUND. 1970. The arrangement of neurosecretory and catecholamine fibres in relation to the pituitary intermedia cells of the skate, *Raja radiata*. *Z. Zellforsch.* 108:81–92, 1970.

184. NAKAI, Y., and A. GORBMAN. Evidence for a doubly innervated secretory unit in the pars intermedia. II. Electron microscopic studies. *Gen. Comp. Endocrinol.* 13:108–116, 1969a.

185. VINCENT, D. S., and T. C. A. KUMAR. Electron-microscopic studies on the pars intermedia of the ferret. *Z. Zellforsch.* 99:185–197, 1969.

186. RODRÍGUEZ, E. M., and J. LA POINTE. Light and electron microscopic study of the pars intermedia of the lizard, *Klauberina riversiana*. *Z. Zellforsch.* 104:1–13, 1970.

187. KOBAYASHI, H., and T. MATSUI. Fine structure of the median eminence and its functional significance. pp. 3–46. In *Frontiers in neuroendocrinology*, W. F. Ganong and L. Martini (Eds.), Oxford University Press, London, 1969.

188. OSHIMA, K., and A. GORBMAN. Pars intermedia: unitary electrical activity regulated by light. *Science* 163:195–197, 1969.

189. OSHIMA, K., and A. GORBMAN. Evidence for a doubly innervated secretory unit in the anuran pars intermedia. I. Electrophysiologic studies. *Gen. Comp. Endocrinol.* 13:98–107, 1969.

190. RODRÍGUEZ, E. M. Ultrastructure of the neurohaemal region of the toad median eminence. *Z. Zellforsch.* 93:182–212, 1969.

191. KASTIN, A. J., and A. V. SCHALLY. MSH activity in pituitaries of rats treated with hypothalamic extracts. *Gen. Comp. Endocrinol.* 7:452–456, 1966.

192. SCHALLY, A. V., and A. J. KASTIN. Purification of a bovine hypothalamic factor which elevates pituitary MSH levels in rats. *Endocrinology* 79:768–772, 1966.

193. RALPH, C. L., and S. SAMPATH. Inhibition by extracts of frog and rat brain of MSH release by frog pars intermedia. *Gen. Comp. Endocrinol.* 7:370–374, 1966.

194. TALEISNIK, S., and M. E. TOMATIS. Melanocyte-stimulating-hormone releasing and inhibiting factors in two hypothalamic extracts. *Endocrinology* 81:819–825, 1967.

195. BERCU, B. B., and H. J. BRINKLEY. Hypothalamic and cerebral cortical inhibition of melanocyte-stimulating hormone secretion in the frog, *Rana pipiens*. *Endocrinology* 80:399–403, 1967.

196. TALEISNIK, S., and R. ORIAS. A melanocyte-stimulating hormone-releasing factor in hypothalamic extracts. *Amer. J. Physiol.* 208:293–296, 1965.

197. VON FRISCH, K. Beiträge zur Physiologie der Pigmentzellen in der Fischhaut. *Pflüg. Arch. Ges. Physiol.* 138:319–387, 1911.

198. BABAK, E. Auf chromatischen Hautfunktion der Amphibien. *Pflüg. Arch. Ges. Physiol.* 131:87–118, 1910.

199. LAURENS, H. The reactions of the melanophores of *Amblystoma* larvae. *J. Exp. Zool.* 18:577–638, 1915.

200. FUCHS, R. F. Der Farbenwechsel und die chromatische Hautfunktion der Tiere. H. Winterstein, *Handb. vergl. Physiol.* 3:1189–1657, 1914.

201. LAURENS, H. The reactions of the melanophores of *Amblystoma* larvae. The supposed influence of the pineal organ. *J. Exp. Zool.* 20:237–261, 1916.

202. McCORD, C. P., and F. P. ALLEN. Evidences associating pineal gland function with alteration in pigmentation. *J. Exp. Zool.* 23:207–224, 1917.

203. SCHARRER, E. Die Lichtempfindlichkeit blinder Elritzen (Untersuchungen über das Zwischenhirn der Fische I). *Z. vergl. Physiol.* 7:1–38, 1928.

204. BORS, O., and W. C. RALSTON. A simple assay of mammalian pineal extracts. *Proc. Soc. Exp. Biol. Med.* 77:807–808, 1951.

205. BAGNARA, J. T. Tail melanophores of *Xenopus* in normal development and regeneration. *Biol. Bull.* 118:1–8, 1960.

206. BAGNARA, J. T. Pineal regulation of the body-lightening reaction in amphibian larvae. *Science* 132:1481–1483, 1960.

207. KELLY, D. E. Pineal organs: photoreception, secretion, and development. *Am. Scientist* 50:597–625, 1962.

208. EAKIN, R. M. Photoreceptors in the amphibian frontal organ. *Proc. Nat. Acad. Sci.* 47:1084–1088, 1961.

209. STEYN, W. Observations on the ultrastructure of the pineal eye. *J. Roy. Micr. Soc.* 79:47–58, 1960.

210. EAKIN, R. M., and J. A. WESTFALL. Fine structure of the retina in the reptilian third eye. *J. Biophys. Biochem. Cytol.* 6:133–134, 1959.

211. EAKIN, R. M., and J. A. WESTFALL. Further observations on the fine structure of the parietal eye of lizards. *Biophys. Biochem. Cytol.* 8:483–501, 1960.

212. KELLY, D. E., and J. C. VAN DE KAMER. Cytological and histochemical investigations on the pineal organ of the adult frog (*Rana esculenta*). *Z. Zellforsch.* 52:618–639, 1960.

213. HAMASAKI, D. I. Properties of the parietal eye of the green iguana. *Vision Res.* 8:591–599, 1968.

214. DODT, E., and E. HEERD. Mode of action of pineal nerve fibers in frogs. *J. Neurophysiol.* 25:405–429, 1962.

215. KLEINE, A. Über die Parietalorgane bei einheimischen und ausländischen Anuran. *Jena L. Med. Naturwiss.* 64:339–376, 1930.

216. STEBBINS, R. C., W. STEYN, and C. PEERS. Results of stirnorganectomy in tadpoles of the African horned frog, *Pyxicephalus delalandi*. *Herpetologica* 16:261–275, 1960.

217. CHARLTON, H. M. The pineal gland and color change in *Xenopus laevis* Daudin. *Gen. Comp. Endocrinol.* 7:384–397, 1966.

218. DODT, E., and M. JACOBSON. Photosensitivity of a localized region of the frog diencephalon. *J. Neurophysiol.* 26:752–758, 1963.

219. BAGNARA, J. T. The pineal and the body-lightening reaction of larval amphibians. *Gen. Comp. Endocrinol.* 3:86–100, 1963.

112. CHAVIN, W. Pituitary hormones in melanogenesis, pp. 63–83. In *Pigment cell biology,* M. Gordon (Ed.). Academic Press, New York, 1959.

113. PICKFORD, G. E. Melanogenesis in *Fundulus heteroclitus. Anat. Rec.* 125:603–604, 1956.

114. PICKFORD, G. E., and B. KOSTO. Hormonal induction of melanogenesis in hypophysectomized killifish (*Fundulus heteroclitus*). *Endocrinology* 61:177–196, 1957.

115. PARKER, G. H. Chemical control of nervous activity. C. Neurohormones in lower vertebrates, pp. 633–656. In *The hormones, Vol. II,* G. Pincus and K. V. Thimann (Eds.), Academic Press, New York, 1950.

116. KLEINHOLZ, L. H. The melanophore-dispersing principle in the hypophysis of *Fundulus heteroclitus. Biol. Bull.* 69:379–390, 1935.

117. KLEINHOLZ, L. H. Studies in reptilian colour changes. II. The pituitary and adrenal glands in the regulation of the melanophores of *Anolis carolinensis. J. Exp. Biol.* 15:474–491, 1938.

118. KLEINHOLZ, L. H. Studies in reptilian colour changes. III. Control of the light phase and behaviour of isolated skin. *J. Exp. Biol.* 15:492–499, 1938.

119. HOROWITZ, S. B. The energy requirements of melanin granule aggregation and dispersion in the melanophores of *Anolis carolinensis. J. Cell. Comp. Physiol.* 51:341–357, 1958.

120. NOBLE, G. K., and H. T. BRADLEY. The relation of the thyroid and the hypophysis to the moulting process in the lizard, *Hemidactylus brookii. Biol. Bull.* 64:289–298, 1933.

121. RAHN, H. The pituitary regulation of melanophores in the rattlesnake. *Biol. Bull.* 80:228–237, 1941.

122. CANELLA, M. F. Note di fisiologia dei cromatofori dei vertebrati pecilotermi, particolarmente dei lacertili. Monitore Zool. Itali. 71:430–480, 1963.

123. YOUNG, J. Z. The photoreceptors of lampreys. II. The functions of the pineal complex. *J. Exp. Biol.* 12:254–270, 1935.

124. EDDY, J. M. P., and R. STRAHAN. The role of the pineal complex in the pigmentary effector system of the lampreys, *Mordacia mordax* (Richardson) and *Geotria australis* Gray. *Gen. Comp. Endocrinol.* 11:528–534, 1968.

125. HOLMBERG, K. Ultrastructure and response to background illumination of the melanophores of the Atlantic hagfish, *Myxine glutinosa* L. *Gen. Comp. Endocrinol.* 10:421–428, 1968.

126. LARSEN, L. O. Effects of hypophysectomy in the cyclostome, *Lampetra fluviatilis* (L) Gray. *Gen. Comp. Endocrinol.* 5:16–30, 1965.

127. PARKER, G. H., and H. PORTER. The control of the dermal melanophores in elasmobranch fishes. *Biol. Bull.* 66:30–37, 1934.

128. GORBMAN, A., and H. A. BERN. *A textbook of comparative endocrinology.* John Wiley and Sons, New York, 1962.

129. KLEINHOLZ, L. H., and H. RAHN. The distribution of intermedin: a new biological method of assay and results under normal and experimental conditions. *Anat. Rec.* 76:157–172, 1940.

130. WITSCHI, E. Vertebrate gonadotrophins. *Mem. Soc. Endocrinol.* 4:149–165, 1955.

131. RALPH, C. L., D. L. GRINWICH, and P. F. HALL. Studies of the melanogenic response of regenerating feathers in the weaver bird: comparison of two species in response to two gonadotrophins. *J. Exp. Zool.* 166:283–288, 1967.

132. GROENENDIJK-HUIJERS, M. M. Development of black down feathering in hybrid chick embryos (Cross New Hampshire ♂ × Light Sussex ♀) after pituitary implantation. *Experientia* 24:501–503, 1968.

133. DuSHANE, G. P. Neural-fold derivatives in the Amphibia. Pigment cells, spinal ganglia, and Rohon-Beard cells. *J. Exp. Zool.* 78:485–503, 1938.

134. STOPPANI, A. O. M., P. F. PIERONI, and A. J. MURRAY. The role of peripheral nervous system in color changes of *Bufo arenarum* Hensel. *J. Exp. Zool.* 31:631–638, 1954.

135. BAGNARA, J. T. Hypophyseal control of guanophores in anuran larvae. *J. Exp. Zool.* 137:265–284, 1958.

136. SAND, A. The comparative physiology of colour response in reptiles and fishes. *Biol. Rev.* 10:361–382, 1935.

137. FORD, E. A contribution to our knowledge of the life-histories of the dogfishes landed at Plymouth. *J. Mar. Biol. Assoc.* 12:468–505, 1921.

138. HOGBEN, L. T. The pigmentary effector system. VII. The chromatic function in elasmobranch fishes. London, *Proc. Roy. Soc.* B, 120:142–158, 1936.

139. VILTER, V. Recherches histologiques et physiologiques sur la fonction pigmentaire des sélaciens. *Bull. Soc. Sci. Arcachon.* 34:65–136, 1937.

140. ZONDEK, B., and H. KROHN. Hormon des Zwischenlappens der Hypophyse (Intermedin). I. Die Rotfarbung der Elritze als Testobjekt. *Klin. Wochsch.* 2:405–408, 1932.

141. ZONDEK, B., and H. KROHN. Hormon des Zwischenlappens der Hypophyse (Intermedin). II. Intermedin im Organismus (Hypophyse, Gehirn). *Klin. Wochsch.* 2:849–853, 1932.

142. ZONDEK, B., and H. KROHN. Hormon des Zwischenlappens der Hypophyse (Intermedin). III. Zur Chemie, Darstellung und Biologie Intermedins. *Klin. Wochsch.* 2:1293–1298, 1932.

143. ZONDEK, B., and H. KROHN. Ein Hormon der Hypophyse. Zwischenlappenhormon (Intermedin). *Naturwissenschaften* 20:134–136, 1932.

144. SAGE, M. Control of prolactin release and its role in color change in the teleost *Gillichthys mirabilis*. *J. Exp. Zool.* 173:121–128, 1970.

145. FRIES, E. F. B. Iridescent white reflecting chromatophores (antaugophores, iridoleucophores) in certain teleost fishes, particularly in *Bathygobius*. *J. Morphol.* 103:203–254, 1958.

146. WARING, H. *Color change mechanisms of cold-blooded vertebrates.* Academic Press, New York, 1963.

147. MITCHELL, G. M., K. DIERST-DAVIES, and F. W. LANDGREBE. *In vivo* antagonism of melanophore-dispersing hormone by melatonin in adult *Xenopus laevis*. *Proc. Roy. Soc. Med.* 62:891–892, 1969.

148. Bradshaw, S. D., and H. Waring. Comparative studies on the biological activity of melanin-dispersing hormone (MDH), pp. 135–152. In *La spécificité zoologique des hormones hypophysaires et de leurs activités*. Centre National de la Recherche Scientifique, Paris, France, 1969.

149. Lerner, A. B., J. D. Case, Y. Takahashi, T. H. Lee, and W. Mori, Isolation of melatonin; the pineal gland factor that lightens melanocytes. *J. Am. Chem. Soc.* 80:2587, 1958.

150. Stoppani, A. O. M. Neuroendocrine mechanism of color change in *Bufo arenarum* Hensel. *Endocrinology* 30:782–786, 1942.

151. Hadley, M. E. Cytophysiological studies on the chromatophores of *Rana pipiens*. Ph.D. Thesis, Brown University, University microfilms, Ann Arbor, Mich, 1966.

152. Etkin, W. On the control of growth and activity of the pars intermedia of the pituitary by the hypothalamus in the tadpole. *J. Exp. Zool.* 86:113–139, 1941.

153. Mazzi, V. Sulla presenza e sul possible significato di fibre neurosecretorie ipothalamo-ipofisaire nel lobo intermedio dell'ipofisi del *Tritone cristato*. *Monitore Zool. Ital.* 62:1–8, 1954.

154. Rowlands, A. The influence of water and light and the pituitary upon the pigmentary system of the common toad (*Bufo bufo bufo*). *J. Exp. Biol.* 31:151–160, 1954.

155. Guardabassi, A. The hypophysis of *Xenopus laevis* Daudin larvae after removal of the anterior hypothalomus. *Gen. Comp. Endocrinol.* 1:348–363, 1961.

156. Etkin, W. Hypothalamic inhibition of the pars intermedia activity in the frog. *Gen. Comp. Endocrinol. Suppl.* 1:70–79, 1962.

157. Etkin, W. Neurosecretory control of the pars intermedia. *Gen. Comp. Endocrinol.* 2:161–169, 1962.

158. Dierst, K. E., and C. L. Ralph. Effect of hypothalamic stimulation on melanophores in the frog. *Gen. Comp. Endocrinol.* 2:347–353, 1962.

159. Jorgensen, C. B., and L. O. Larsen. Nature of the nervous control of the pars intermedia function in amphibians: rate of functional recovery after denervation. *Gen. Comp. Endocrinol.* 3:468–472, 1963.

160. Dierst-Davies, K., C. L. Ralph, and J. L. Pechersky. Effects of pharmacological agents on the hypothalamus of *Rana pipiens* in relation to the control of skin melanophores. *Gen. Comp. Endocrinol.* 6:409–419, 1966.

161. Sawyer, C. H. Cholinergic stimulation of the release of melanophore hormone by the hypophysis in salamander larvae. *J. Exp. Zool.* 106:145–179, 1947.

162. Enemar, A., and B. Falck. On the presence of adrenergic nerves in the pars intermedia of the frog, *Rana temporaria. Gen. Comp. Endocrinol.* 5:577–583, 1965.

163. Hillarp, N. A., and D. Jacobsohn. Über die Innervation der Adenohypophyse und ihre Beziehungen zur gonadotropen Hypophysenfunktion. *Kungl. Fysiogr. Sällsk. Handl. N.F.* 54:1–25, 1943.

164. BJORKLUND, A., A. ENEMAR, and B. FALCK. Monoamines in the hypothalamo-hypophyseal system of the mouse with special reference to the ontogenetic aspects. *Z. Zellforsch.* 89:590–607, 1968.

165. ITURRIZA, F. C. Electron-microscopic study of the pars intermedia of the pituitary of the toad, *Bufo arenarum. Gen. Comp. Endocrinol.* 4:492–502, 1964.

166. ITURRIZA, F. C. Monoamines and control of pars intermedia in the toad pituitary. *Gen. Comp. Endocrinol.* 6:19–25, 1966.

167. ITURRIZA, F. C. Monoamines in the neurointermediate lobe of the pituitary of the Argentinian eel. *Naturwissenschaften* 21:565, 1967.

168. ITURRIZA, F. C. The secretion of intermedin in autotransplants of pars intermedia growing in the anterior chamber of intact and sympathectomized eyes of the toad. *Neuroendocrinology* 2:11–18, 1967.

169. ITURRIZA, F. C. Further evidences for the blocking effect of catecholamines on the secretion of melanocyte-stimulating hormone in toads. *Gen. Comp. Endocrinol.* 12:417–426, 1969.

170. MASUR, S. K. Fine structure of the autotransplanted pituitary of the red eft, *Notophthalmus viridescens. Gen. Comp. Endocrinol.* 12:12–32, 1969.

171. ITO, T. Experimental studies on the hypothalamic control of the pars intermedia activity of the frog, *Rana nigromaculata. Neuroendocrinology* 3:25–33, 1968.

172. KASTIN, A. J., and G. T. ROSS. Melanocyte-stimulating hormone activity in pituitaries of frogs with hypothalamic lesions. *Endocrinology* 77:45–48, 1965.

173. PEHLEMANN, F. W. Experimentelle Beeinflussung der Melanophorenverteilung von *Xenopus laevis*-larven. *Zool. Anz. Suppl.* 29:571–580, 1967.

174. SALAND, L. C. Ultrastructure of the pars intermedia in relation to hypothalamic control of hormone release. *Nueroendocrinology* 3:72–88, 1968.

175. PEHLEMANN, F. W. Ultrastructure and innervation of the pars intermedia of the pituitary of *Xenopus laevis. Gen. Comp. Endocrinol.* 9:481, 1967.

176. COHEN, A. G. Observations on the pars intermedia of *Xenopus laevis. Nature* 215:55–56, 1967.

177. BURGERS, A. C. J., K. IMAI, and G. J. VAN OORDT. The amount of melanophore-stimulating hormone in single pituitary glands of *Xenopus laevis* kept under various conditions. *Gen. Comp. Endocrinol.* 3:53–57, 1963.

178. ORTMAN, R. Cytochemical study of the physiological activities in the pars intermedia of *Rana pipiens. Anat. Rec.* 119:1–9, 1954.

179. RUST, C. C., and R. K. MEYER. Effect of pituitary autographs on hair color in the short-tailed weasel *Gen. Comp. Endocrinol.* 11:548–551, 1968.

180. HOWE, A., and A. J. THODY. The effect of hypothalamic lesions on the melanocyte-stimulating hormone content and histology of the pars intermedia of the rat pituitary gland. *J. Physiol.* 201:25P–26P, 1969.

181. BAL, H., and P. G. SMELIK. Effect of hypothalamic lesions of MSH content of the intermediate lobe of the pituitary gland in the rat. *Experientia* 23:759–760, 1967.

182. KNOWLES, F. 1965. Evidence for a dual control, by neurosecretion, of hormone synthesis and hormone release in the pituitary of the dogfish, *Scylliorhinus stellaris*. London, *Phil. Trans. Roy. Soc.* B, 249:435–456.

183. MEURLING, P., and A. BJÖRKLUND. 1970. The arrangement of neurosecretory and catecholamine fibres in relation to the pituitary intermedia cells of the skate, *Raja radiata*. Z. *Zellforsch.* 108:81–92, 1970.

184. NAKAI, Y., and A. GORBMAN. Evidence for a doubly innervated secretory unit in the pars intermedia. II. Electron microscopic studies. *Gen. Comp. Endocrinol.* 13:108–116, 1969a.

185. VINCENT, D. S., and T. C. A. KUMAR. Electron-microscopic studies on the pars intermedia of the ferret. Z. *Zellforsch.* 99:185–197, 1969.

186. RODRÍGUEZ, E. M., and J. LA POINTE. Light and electron microscopic study of the pars intermedia of the lizard, *Klauberina riversiana*. Z. *Zellforsch.* 104:1–13, 1970.

187. KOBAYASHI, H., and T. MATSUI. Fine structure of the median eminence and its functional significance. pp. 3–46. In *Frontiers in neuroendocrinology*, W. F. Ganong and L. Martini (Eds.), Oxford University Press, London, 1969.

188. OSHIMA, K., and A. GORBMAN. Pars intermedia: unitary electrical activity regulated by light. *Science* 163:195–197, 1969.

189. OSHIMA, K., and A. GORBMAN. Evidence for a doubly innervated secretory unit in the anuran pars intermedia. I. Electrophysiologic studies. *Gen. Comp. Endocrinol.* 13:98–107, 1969.

190. RODRÍGUEZ, E. M. Ultrastructure of the neurohaemal region of the toad median eminence. Z. *Zellforsch.* 93:182–212, 1969.

191. KASTIN, A. J., and A. V. SCHALLY. MSH activity in pituitaries of rats treated with hypothalamic extracts. *Gen. Comp. Endocrinol.* 7:452–456, 1966.

192. SCHALLY, A. V., and A. J. KASTIN. Purification of a bovine hypothalamic factor which elevates pituitary MSH levels in rats. *Endocrinology* 79:768–772, 1966.

193. RALPH, C. L., and S. SAMPATH. Inhibition by extracts of frog and rat brain of MSH release by frog pars intermedia. *Gen. Comp. Endocrinol.* 7:370–374, 1966.

194. TALEISNIK, S., and M. E. TOMATIS. Melanocyte-stimulating-hormone releasing and inhibiting factors in two hypothalamic extracts. *Endocrinology* 81:819–825, 1967.

195. BERCU, B. B., and H. J. BRINKLEY. Hypothalamic and cerebral cortical inhibition of melanocyte-stimulating hormone secretion in the frog, *Rana pipiens*. *Endocrinology* 80:399–403, 1967.

196. TALEISNIK, S., and R. ORIAS. A melanocyte-stimulating hormone-releasing factor in hypothalamic extracts. *Amer. J. Physiol.* 208:293–296, 1965.

197. VON FRISCH, K. Beiträge zur Physiologie der Pigmentzellen in der Fischhaut. *Pflüg. Arch. Ges. Physiol.* 138:319–387, 1911.

198. BABAK, E. Auf chromatischen Hautfunktion der Amphibien. *Pflüg. Arch. Ges. Physiol.* 131:87–118, 1910.

199. LAURENS, H. The reactions of the melanophores of *Amblystoma* larvae. *J. Exp. Zool.* 18:577–638, 1915.

200. FUCHS, R. F. Der Farbenwechsel und die chromatische Hautfunktion der Tiere. H. Winterstein, *Handb. vergl. Physiol.* 3:1189–1657, 1914.

201. LAURENS, H. The reactions of the melanophores of *Amblystoma* larvae. The supposed influence of the pineal organ. *J. Exp. Zool.* 20:237–261, 1916.

202. McCORD, C. P., and F. P. ALLEN. Evidences associating pineal gland function with alteration in pigmentation. *J. Exp. Zool.* 23:207–224, 1917.

203. SCHARRER, E. Die Lichtempfindlichkeit blinder Elritzen (Untersuchungen über das Zwischenhirn der Fische I). *Z. vergl. Physiol.* 7:1–38, 1928.

204. BORS, O., and W. C. RALSTON. A simple assay of mammalian pineal extracts. *Proc. Soc. Exp. Biol. Med.* 77:807–808, 1951.

205. BAGNARA, J. T. Tail melanophores of *Xenopus* in normal development and regeneration. *Biol. Bull.* 118:1–8, 1960.

206. BAGNARA, J. T. Pineal regulation of the body-lightening reaction in amphibian larvae. *Science* 132:1481–1483, 1960.

207. KELLY, D. E. Pineal organs: photoreception, secretion, and development. *Am. Scientist* 50:597–625, 1962.

208. EAKIN, R. M. Photoreceptors in the amphibian frontal organ. *Proc. Nat. Acad. Sci.* 47:1084–1088, 1961.

209. STEYN, W. Observations on the ultrastructure of the pineal eye. *J. Roy. Micr. Soc.* 79:47–58, 1960.

210. EAKIN, R. M., and J. A. WESTFALL. Fine structure of the retina in the reptilian third eye. *J. Biophys. Biochem. Cytol.* 6:133–134, 1959.

211. EAKIN, R. M., and J. A. WESTFALL. Further observations on the fine structure of the parietal eye of lizards. *Biophys. Biochem. Cytol.* 8:483–501, 1960.

212. KELLY, D. E., and J. C. VAN DE KAMER. Cytological and histochemical investigations on the pineal organ of the adult frog (*Rana esculenta*). *Z. Zellforsch.* 52:618–639, 1960.

213. HAMASAKI, D. I. Properties of the parietal eye of the green iguana. *Vision Res.* 8:591–599, 1968.

214. DODT, E., and E. HEERD. Mode of action of pineal nerve fibers in frogs. *J. Neurophysiol.* 25:405–429, 1962.

215. KLEINE, A. Über die Parietalorgane bei einheimischen und ausländischen Anuran. *Jena L. Med. Naturwiss.* 64:339–376, 1930.

216. STEBBINS, R. C., W. STEYN, and C. PEERS. Results of stirnorganectomy in tadpoles of the African horned frog, *Pyxicephalus delalandi*. *Herpetologica* 16:261–275, 1960.

217. CHARLTON, H. M. The pineal gland and color change in *Xenopus laevis* Daudin. *Gen. Comp. Endocrinol.* 7:384–397, 1966.

218. DODT, E., and M. JACOBSON. Photosensitivity of a localized region of the frog diencephalon. *J. Neurophysiol.* 26:752–758, 1963.

219. BAGNARA, J. T. The pineal and the body-lightening reaction of larval amphibians. *Gen. Comp. Endocrinol.* 3:86–100, 1963.

220. BRICK, I. Relationship of the pineal to the pituitary-melanophore effector system in *Amblystoma opacum. Anat. Rec.* 142:299, 1962.

221. QUAY, W. B., and J. T. BAGNARA. Relative potencies of indolic and related compounds in the body-lightening reaction of larval *Xenopus. Arch. Intern. Pharmacodyn.* 150:137–143, 1964.

222. BURGERS, A. C. J., and G. J. VAN OORDT. Regulation of pigment migration in the amphibian melanophore. *Gen. Comp. Endocrinol. Suppl.* 1:99–109, 1962.

223. HOOKER, D. The reactions of light and darkness on the melanophores of frog tadpoles. *Science* 39:473, 1914.

224. BAGNARA, J. T. Independent actions of pineal and hypophysis in the regulation of chromatophores of anuran larvae. *Gen. Comp. Endocrinol.* 4:299–303, 1964.

225. BAGNARA, J. T., M. E. HADLEY, and J. D. TAYLOR. Regulation of bright-colored pigmentation of amphibians. *Gen. Comp. Endocrinol. Suppl.* 2:425–438, 1969.

226. RUST, C. C., and R. K. MEYER. Hair color, molt, and testis size in male, short-tailed weasels treated with melatonin. *Science* 165:921–922, 1969.

227. KASTIN, A. J., A. V. SCHALLY, S. VIOSCA, L. BARRETT, and T. W. REDDING. MSH activity in rat pituitaries after constant illumination. *Neuroendocrinology* 2:257, 1967.

228. KASTIN, A. J., and A. V. SCHALLY. MSH release in mammals. pp. 215–224. In *Pigmentation: its genesis and control,* V. Riley (Ed.), Appleton-Century-Crofts, New York, 1972.

229. KASTIN, A. J., S. VIOSCA, and A. V. SCHALLY. Assay of mammalian MSH-release regulating factor(s). pp. 171–183. In *Hypophysiotropic hormones of the hypothalamus: assay and chemistry,* J. Meites (Ed.), The Williams and Wilkins Company, Baltimore, 1970.

230. SNELL, R. S. Effect of melatonin on mammalian epidermal melanocytes. *J. Invest. Derm.* 44:273–275, 1965.

231. LERNER, A. B. Hormones and skin color. *Sci. Amer.* 204 (July): 98–108, 1961.

232. REAMS, W. M., R. E. SHERVETTE, and W. H. DORMAN. Refractoriness of mouse dermal melanocytes to hormones. J. Invest. Derm. 50:338–339, 1968.

233. WYMAN, L. C. The reactions of the melanophores of embryonic and larval *Fundulus* to certain chemical substances. *J. Exp. Zool.* 40:161–180, 1924.

234. WYMAN, L. C. Blood and nerve as controlling agents in the movements of melanophores. *J. Exp. Zool.* 39:73–132, 1924.

235. HEWER, H. R. Studies in colour changes of fish. I. The action of certain endocrine secretions in the minnow. *J. Exp. Biol.* 3:123–140, 1926.

236. NICHOLS, J., C. SCHNEEBECK, and M. E. HADLEY. Comparative *in vivo* response of embryonic, larval, and adult melanophores of *Fundulus heteroclitus* to melatonin. *Amer. Zool.* 6:576, 1966.

237. FAIN, W. B., and M. E. HADLEY. *In vitro* response of melanophores of *Fundulus heteroclitus* to melatonin, adrenaline, and noradrenaline. *Amer. Zool.* 6:596, 1966.

238. REED, B. L. The control of circadian pigment changes in the pencil fish: a proposed role for melatonin. *Life Sci.* 7:961–973, 1968.

239. HAFEEZ, M. A. Effect of melatonin on body coloration and spontaneous swimming activity in rainbow trout, *Salmo gairdneri. Comp. Biochem. Physiol.* 36:639–656, 1970.

240. PARKER, G. H., and A. J. LANCHER. The response of *Fundulus* to white, black and darkness. *Amer. J. Physiol.* 61:548–550, 1922.

241. MATTHEWS, S. A. Color changes in hypophysectomized *Fundulus. Biol. Bull.* 64:315–320, 1933.

242. ABRAMOWITZ, A. A. Physiology of the melanophore system in the catfish, *Ameiurus. Biol. Bull.* 71:259–281, 1936.

243. PANG, P. K. T. The effect of pinealectomy on adult male killifish, *Fundulus heteroclitus. Amer. Zool.* 7:715, 1967.

244. HAFEEZ, M. A., and W. B. QUAY. The role of the pineal organ in the control of phototaxis and body coloration in rainbow trout (*Salmo gairdneri*, Richardson). *Z. vergl. Physiol.* 68:403–416, 1970.

245. OKSCHE, A., and H. KIRSHSTEIN. Die Ultrastruktur der Sinneszellen im Pinealorgan von *Phoxinus laevis* L. *Z. Zellforsch.* 78:151–161, 1967.

246. BREDER, C. M., JR., and P. RASQUIN. A preliminary report on the role of the pineal organ in the control of pigment cells and light reactions in recent teleost fishes, *Science* 111:10–12, 1950.

247. PARKER, G. H. The colour changes in lizards, particularly in *Phrynosoma. J. Exp. Biol.* 15:48–73, 1938.

248. CARLTON, F. C. The color changes in the skin of the so-called Florida chameleon, *Anolis carolinensis Cuv. Proc. Am. Acad. Arts Sci.* 39:259–276, 1903.

249. RALPH, C. L., L. HEDLUND, and W. A. MURPHY. Diurnal cycles of melatonin in bird pineal bodies. *Comp. Biochem. Physiol.* 22:591–599, 1967.

250. BAGNARA, J. T., and M. E. HADLEY. Endocrinology of the amphibian pineal. *Am. Zool.* 10:201–216, 1970.

251. BOGENSCHÜTZ, H. Über den Farbwechsel von *Rana esculenta* nach Epiphysektomie. *Experientia* 23:967–968, 1967.

252. BRÜCKE, E. Untersuchungen über den Farbwechsel des afrikanischen Chamaleons. *Denkschr. Akad. Wiss. Wien.* 4:179–210, 1852.

253. POUCHET, G. Des changements de coloration sous l'influence des nerfs. *J. Anat. Physiol.* 12:1–90, 113–65, 1876.

254. BALLOWITZ, E. Die Innervation der Chromatophoren. *Verg. Anat. Ges. Jena* 7:71–76, 1893.

255. SPAETH, R. A., and H. G. BARBOUR. The action of epinephrin and ergotoxin upon single physiologically isolated cells. *J. Pharmacol. Exp. Ther.* 9:431–440, 1917.

256. SCHELINE, R. R. Adrenergic mechanisms in fish: chromatophore pigment concentration in the cuckoo wrasse *Labris ossifagus* L. *Comp. Biochem. Physiol.* 9:215–227, 1963.

257. Falck, B., J. Muntzing, and A. M. Rosengren. Adrenergic nerves to the dermal melanophores of the rainbow trout, *Salmo gairdneri. Z. Zellforsch.* 99:430–434, 1969.

258. Jacobowitz, D. M., and A. M. Laties. Direct adrenergic innervation of a teleost melanophore. *Anat. Rec.,* 162:501–504, 1968.

259. Bikle, D., L. G. Tilney, and K. R. Porter. Microtubules and pigment migration in melanophores of *Fundulus heteroclitus* L. *Protoplasma* 61:322–345, 1966.

260. Fujii, R. Correlation between fine structure and activity in fish melanophore. pp. 114–123. In *Structure and control of the melanocyte,* G. Della Porta and O. Mühlbock (Eds.), Springer-Verlag, New York, 1966.

261. Fujii, R., and R. Novales. The nervous mechanism controlling pigment aggregation in *Fundulus melanophores. Comp. Biochem. Physiol.* 29:109–124, 1969.

262. Ruffin, N. E., B. L. Reed, and B. C. Finnin. Pharmacological studies on teleost melanophores. *Europ. J. Pharmacol.* 8:114–118, 1969.

263. Abbott, F. S. The effects of certain drugs and biogenic substances on the melanophores of *Fundulus heteroclitus* L. *Canad. J. Zool.* 46:1149–1161, 1968.

264. Hogben L. T., and L. Mirvish. The pigmentary effector system. V. The nervous control of excitement pallor in reptiles. *J. Exp. Biol.* 5:295–308, 1928.

265. Redfield, A. C. The coordination of chromatophores by hormones. *Science* 43:580–581, 1916.

266. Redfield, A. C. The physiology of the melanophores of the horned toad *Phrynosoma. J. Exp. Zool.* 26:275–333, 1918.

267. Kleinholz, L. H. Studies in reptilian color changes I. A preliminary report. *Proc. Nat. Acad. Sci.* 22:454–456, 1936.

268. Wykes, V. Observations on pigmentary coordination in elasmobranchs. *J. Exp. Biol.* 13:460–466, 1936.

269. Young, J. Z. The autonomic nervous system of selachians. *Quart. J. Micr. Sci.* 75:571–624, 1933.

270. Parker, G. H. Color changes in *Mustelus* and other elasmobranch fishes. *J. Exp. Zool.* 89:451–473, 1942.

271. Abramowitz, A. A. The pituitary control of chromatophores in the dogfish. *Amer. Nat.* 73:208–218, 1939.

272. Snell, R. S., and S. Kulovich. Nerve stimulation and the movement of melanin granules in the pigment cells of the frog's web. *J. Invest. Derm.* 58:438–443, 1967.

273. Falck, B., S. Jacobsson, H. Olivecrona, and H. Rorsman. Pigmented nevi and malignant melanomas as studied with a specific fluorescence method. *Science* 149:439–440, 1965.

274. Falck, B., W. C. Jacobson, H. Olivecrona, and H. Rorsman. Fluorescent dopa reaction of nevi and melanomas. *Arch. Derm.* 94:363–369, 1966.

275. SERRI, F., and D. CERIMELE. Connections between nerve fibers and dendritic cells in the skin of the fetus. *Advan. Biol. Skin* 8:31–39, 1967.

276. LERNER, A. B. Vitiligo. *J. Invest. Derm.* 32:285–310, 1959.

277. CHANCO-TURNER, M. L., and A. B. LERNER. Physiologic changes in vitiligo. *Arch. Derm.* 91:390–396, 1965.

278. SPAETH, R. A. Evidence proving the melanophore to be disguised type of smooth muscle cell. *J. Exp. Zool.* 20:193–215, 1916.

279. WRIGHT, P. A. Physiological responses of frog melanophores *in vitro*. *Physiol. Zool.* 28:204–218, 1955.

280. BURGERS, A. C. J. Biological aspects of pigment cell research, pp. 6–16. In *Structure and control of the melanocyte*. G. Della Porta and O. Mühlbock (Eds.). Springer-Verlag, New York, 1966.

281. NOVALES, R. R., and W. J. DAVIS. Cellular aspects of the control of physiological color changes in amphibians. *Amer. Zool.* 9:479–488, 1969.

282. BURGERS, A. C. J., TH. A. C. BOSCHMAN, and J. C. VAN DE KAMER. Excitement darkening and the effect of adrenaline on the melanophores of *Xenopus laevis*. *Acta Endocrinol.* 14:72–82, 1953.

283. HADLEY, M. E., and J. M. GOLDMAN. The physiological regulation of the amphibian iridophore. pp. 225–246. In *Pigmentation: its genesis and control*, V. Riley (Ed.), Appleton-Century-Crofts, New York, 1972.

284. MÖLLER, H., and A. B. LERNER. Melanocyte-stimulating hormone inhibition by acetylcholine and noradrenaline in the frog skin bioassay. *Acta Endocrinol.* 51:149–160, 1966.

285. PARKER, G. H. Effects of acetyl choline on chromatophores. *Proc. Nat. Acad. Sci.* 17:596–597, 1931.

286. PARKER, G. H. Acetyl choline and chromatophores. *Proc. Nat. Acad. Sci.* 20:596–599, 1934.

287. WRIGHT, M. R., and A. B. LERNER. On the movement of pigment granules in frog melanocytes. *Endocrinology* 66:599–609, 1960.

288. HIMES, P. J., and M. E. HADLEY. *In vitro* effects of steroid hormones on vertebrate melanophores. *J. Invest. Derm.* 57:337–342, 1971.

289. ODELL, W. D., and G. T. ROSS. Lack of effect of adrenal cortical steroids on the pigmentary response of intact frog. *Endocrinology* 73:647–649, 1963.

290. BISCHITZ, P. G., and R. S. SNELL. A study of the effect of ovariectomy, estrogen and progesterone on the melanocytes and melanin in the skin of the female guinea-pig. *J. Endocrinol.* 20:312–319, 1960.

291. SNELL, R. S. The pigmentary changes occurring in the breast skin during pregnancy and following estrogen treatment. *J. Invest. Derm.* 43:181–186, 1964.

292. McGILL, T. E., and G. R. TUCKER. Coat "color" difference between castrated and intact male mice. *Experientia* 23:574, 1967.

293. KUPPERMANN, H. S. Hormone control of a dimorphic pigmentation area in the golden hamster (*Cricetus auratus*). *Anat. Rec.*, 88:442, 1944.

294. MILLS, T. M., and E. SPAZIANI. Hormonal control of melanin pigmentation in scrotal skin of rat. *Exp. Cell Res.* 44:13–22, 1966.

295. WILSON, M. J., and E. SPAZIANI. Regulation by testosterone of pigmentation in the scrotal skin of the rat. *Am. Zool.* 9:1090–1091, 1969.

296. LERNER, A. B., K. SHIZUME, and I. BUNDING. The mechanism of endocrine control of melanin pigmentation. *J. Clin. Endocrinol. Metab.* 14:1463–1490, 1954.

297. WILLIER, B. H. Cells, feathers, and colors. *Bios* 23:109–125, 1952.

298. WITSCHI, E. The quantitative determination of follicle stimulating and luteinizing hormones in mammalian pituitaries and a discussion of the gonadotropic quotient, F/L. *Endocrinology* 27:437–446, 1940.

299. SEGAL, S. J. Response of weaver finch to chorionic gonadotropin and hypophysial luteinizing hormone. *Science* 126:1242–1243, 1957.

300. WITSCHI, E. Effect of gonadotrophic and oestrogenic hormones on regenerating feathers of weaver finches (*Pyromelana franciscana*). *Proc. Soc. Exp. Biol. Med.* 35:484–489, 1936.

301. HALL, P. F., C. L. RALPH, and D. L. GRINWICH. On the locus of action of interstitial cell-stimulating hormone (ICSH or LH) on feather pigmentation of African weaver birds. *Gen. Comp. Endocrinol.* 5:552–557, 1965.

302. PICKFORD, G. E., and J. W. ATZ. The chromatophore hormones of the pituitary. pp. 32–59. In *The physiology of the pituitary gland of fishes*. New York Zoological Society, New York, 1957.

303. WORONZOWA, M. A. Analyse der weissen Fleckung bei Amblystomen. *Biol. Zbl.* 52:676–684, 1932.

304. CHANG, C. Y. Thyroxine effect on melanophore contraction in *Xenopus laevis*. *Science* 126:121–122, 1957.

305. WRIGHT, M. R., and A. B. LERNER. Action of thyroxine analogues on frog melanocytes. *Nature* 185:169–170, 1960.

306. LAROCHE, G., and C. P. LEBLOND. Effect of thyroid preparations and iodine on Salmonidae. *Endocrinology* 51:524–545, 1952.

307. ROBERTSON, O. H. Production of the silvery smolt in rainbow trout by intramuscular injection of mammalian thyroid extract and thyrotropic hormone. *J. Exp. Zool.* 110:337–355, 1949.

308. LAROCHE, G., A. N. WOODALL, C. L. JOHNSON, and J. E. HALVER. Thyroid function in the rainbow trout (*Salmo gairdnerii Rich.*). II. Effects of thyroidectomy on the development of young fish. *Gen. Comp. Endocrinol.* 6:249–266, 1966.

309. FINGERMAN, M. Cellular aspects of the control of physiological color changes in crustaceans. *Am. Zool.* 9:443–452, 1969.

310. FINGERMAN, M. Chromatophores. *Physiol. Rev.* 45:296–339, 1965.

311. FINGERMAN, M. Comparative physiology: chromatophores. *Ann. Rev. Physiol.* 32:345–372, 1970.

312. BROWN, F. A., JR. Chromatophores and color change. pp. 502–537. In *Comparative animal physiology*. W. B. Saunders Company, Philadelphia, 1961.

313. DARWIN, C. The voyage of the beagle. pp. 7–8. *The Natural History Library*, Anchor Books. Doubleday and Company, Garden City, N.Y.

314. FLOREY, E. Ultrastructure and function of cephalopod chromatophores. *Am. Zool.* 9:429–442, 1969.

315. FLOREY, E., and M. E. KRIEBEL. Electrical and mechanical responses of chromatophore muscle fibers of the squid, *Loligo opalescens,* to nerve stimulation and drugs. *Z. vergl. Physiol.* 65:98–130, 1969.

316. WEBER, W. Muptiple Innervation der Chromatophorenmuskelzellen von *Loligo vulgaris. Z. Zellforsch.* 92:367–376, 1968.

317. WEBER, W. Zur Ultrastruktur der Chromatophorenmuskelzellen von *Loligo vulgaris. Z. Zellforsch.* 108:446–456, 1970.

318. BROWN, F. A., JR., M. FINGERMAN, M. I. SANDEEN, and H. M. WEBB. Persistent diurnal and tidal rhythms of color change in the fiddler crab, *Uca pugnax. J. Exp. Zool.* 123:29–60, 1953.

319. POUCHET, G. Sur les changements de coloration que présentent certains poissons et certains crustacés. *C. R. Soc. Biol.* (*Paris*) 24:63–65, 1874.

320. PERKINS, E. B. Color changes in crustaceans, especially in *Palaemonetes. J. Exp. Zool.* 11:703–712, 1928.

321. KOLLER, G. Über Chromatophorensystem, Farbensinn und Farbwechsel bei *Crangon vulgaris. Z. vergl. Physiol.* 5:191–246, 1927.

322. SHIBLEY, G. A. Eyestalk function in chromatophore control in a crab, *Cancer magister. Physiol. Zool.* 41:268–279, 1968.

323. ABRAMOWITZ, A. A. The chromatophorotropic hormone of the Crustacea: standardization, properties and physiology of the eyestalk glands. *Biol. Bull.* 72:344–365, 1937.

324. BROWN, F. A., JR. The chemical nature of the pigments and the transformations responsible for color changes in *Palaemonetes. Biol. Bull.* 67:365–380, 1934.

325. BROWN, F. A., JR., and H. H. SCUDAMORE. Differentiation of two principles from the crustacean sinus-gland. *J. Cell. Comp. Physiol.* 15:103–119, 1940.

326. FINGERMAN, M. Neurosecretory control of pigmentary effectors in Crustaceans. *Am. Zool.* 6:169–179, 1966.

327. KLEINHOLZ, L. H. Separation and purification of crustacean eyestalk hormones. *Amer. Zool.* 6:161–167, 1966.

328. KLEINHOLZ, L. H. Pigmentary-effector hormones from the neurosecretory system of crustacean eyestalks. *Gen. Comp. Endocrinol.* 14:578–588, 1970.

329. BROWN, F. A., JR. Color changes in *Palaemonetes. J. Morphol.* 57:317–334, 1935.

330. RAO, K. R. Chromatophorotropic activities of the saline extracts of different portions of the nervous system of the crab, *Ocypode macrocera. Broteria* 37:59–70, 1968.

331. CHASSARD-BOUCHAUD, C. L'adaptation chromatique chez les Natantia (Crustacés Décapodes). *Cahiers Biol. Marine* 6:469–576, 1965.

332. AOTO, T. Rate of production and release of chromatophorotropins in crustaceans. *Gen. Comp. Endocrinol. Suppl.* 2:459–467, 1969.

333. RAO, K. R., and M. FINGERMAN. Action of biogenic amines on crustacean chromatophores, I. Differential effect of certain indolealkylamines on melanophores of the crabs *Uca pugilator* and *Carcinus maenas*. *Experientia* 26:383–384, 1970.

334. RAO, K. R., and M. FINGERMAN. Action of biogenic amines on crustacean chromatophores, II. Analysis of the responses of erythrophores in the fiddler crab, *Uca pugilator*, to indolealkylamines and an eyestalk hormone. *Comp. Gen. Pharmacol.* I:117–126, 1970.

335. LEE, W. L. Color change and the ecology of the marine isopod *Idothea (Pentidotea) montereyensis* Maloney, 1933. *Ecology* 47:930–941, 1966.

336. LEE, W. L. Pigmentation of the marine isopod *Idothea montereyensis*. *Comp. Biochem. Physiol.* 18:17–36, 1966.

337. KLEINHOLZ, L. H. Crustacean eyestalk-hormone and retinal pigment migration. *Biol. Bull.* 70:159–184, 1936.

338. NICOL, J. A. C. *The biology of marine animals*. John Wiley and Sons, New York, 1967.

339. WELSH, J. H. Diurnal rhythm of the distal pigment cells in the eyes of certain crustaceans. *Proc. Nat. Acad. Sci.* 16:386–395, 1930.

340. BENNITT, R. Diurnal rhythm in the proximal pigment cells of the crayfish retina. *Physiol. Zool.* 5:65–69, 1932.

341. HENKES, H. E. Retinomotor and diurnal rhythm in crustaceans. *J. Exp. Biol.* 29:178–191, 1952.

342. ALI, M. A. The ocular structure, retinomotor, and photo-behavioral response of juvenile Pacific Salmon. *Can. J. Zool.* 37:965–996, 1959.

343. BIGNEY, A. J. The effect of adrenin on the pigment migration in the melanophores of the skin and in the pigment cells of the retina of the frog. *J. Exp. Zool.* 27:391–396, 1919.

344. KRAUS-RUPPERT, R., and F. LEMBECK. Die Wirkung von Melatonin auf die Pigmentzellen der Retina von Fröschen. *Pflüger's Arch. Ges. Physiol.* 284:160–168, 1965.

345. LEVITA, B. New method of spectrophotometric and colorimetric investigations on insect integuments *in vivo*. *Zeiss Information No.* 68:45–50, 1969.

346. KEY, K. H. L., and M. F. DAY. A temperature controlled physiological colour response in the grasshopper *Koscuscola tristis*, Sjöst (Orthoptera Acrididae). *Aust. J. Zool.* 2:309–339, 1954.

347. GIERSBERG, H. Color change in *Carausius*; influence of humidity. *Z. vergl. Physiol.* 7:657–695, 1928.

348. DUPONT-RAABE, M. Chromatophorotropins of insects. *C. R. Acad. Sci.* 228:130–132, 1949.

349. KARLSON, P., and C. E. SEKERIS. Zum Tyrosinstoffwechsel der Insekten. IX. Kontrolle des Tyrosinstoffwechsels durch Ecdyson. *Biochim. Biophys. Acta* 63:489–495, 1962.

350. FRAENKEL, G., and C. HSIAO. Hormonal and nervous control of tanning in the fly. *Science* 138:27–29, 1963.

351. Fraenkel, G., and C. Hsiao. Bursicon, a hormone which mediates tanning of the cuticle in the adult fly and other insects. *J. Insect. Physiol.* 11:513–556, 1965.

352. Seligman, M., S. Friedman, and G. Fraenkel. Bursicon mediation of tyrosine hydroxylation during tanning of the adult cuticle of the fly, *Scarcophaga bullata*. *J. Insect. Physiol.* 15:553–561, 1969.

353. Mills, R. R., R. B. Mathur, and A. A. Guerra. Studies on the hormonal control of tanning in the American cockroach. 1. Release of an activation factor from the terminal abdominal ganglion. *J. Insect. Physiol.* 11:1047–1053, 1965.

354. Uvarov, B. P. A. revision of the genus *Locusta* L. [= *Pachytylus Fiab.*] with a new theory as to the periodicity and migrations of locusts. *Brit. Ent. Res.* 12:135–163, 1921.

355. Faure, J. C. Color change in locusts. *Bull. Ent. Res.* 23:293–405, 1932.

356. Nolte, D. J. A pheromone for melanization of locusts. *Nature* 200:660–661, 1963.

357. Watt, W. B. Adaptive significance of pigment polymorphisms in *Colias* butterflies, II. Thermoregulation and photoperiodically controlled melanin variation in *Colias eurytheme*. *Proc. Nat. Acad. Sci.* 63:767–774, 1969.

358. Watt, W. B. Adaptive significance of pigment polymorphisms in *Colias* butterflies. I. Variation of melanin pigment in relation to thermoregulation. *Evolution* 22:437–458, 1968.

359. Kettlewell, H. B. D. The phenomenon of industrial melanism in Lepidoptera. *Ann. Rev. Entomol.* 6:245–262, 1961.

360. Clapiarede, E. Les Annélides chétopodes du golfe de Napoles. Supplément. Genève et Bâle, 1870.

361. Smith, R. I. Nervous control of chromatophores in the leech *Placobdella parasitica*. *Physiol. Zool.* 15:410–417, 1942.

362. Wells, G. P. Colour response in a leech. *Nature* 129:686–687, 1932.

363. Millott, N. Colour change in the echinoid, *Diadema antillarum* Philippi. *Nature* (London) 170:325–326, 1952.

364. Yoshida, M. On the light response of the chromatophore of the sea-urchin, *Diadema setosum*. *J. Exp. Biol.* 33:119–123, 1956.

365. von Üexkull, J. Vergleichend sinnesphysiologische Untersuchungen. II. Der Schatten als Reiz für *Centrostephanus longispinus*. *Z. Biol.* 34:319–339, 1896.

366. Kleinholz, L. H. Color changes in echinoderms, *Pubbl. Staz. Zool. Napoli* 17:53, 1938.

367. Parker, G. H. The color changes in the sea-urchin *Arbacia*. *Proc. Nat. Acad. Sci.* 17:594–596, 1931.

368. Dambach, M., u. F. Jochum. Zum Verlauf der Pigmentausbreitung beim Farbwechsel des Seeigels *Centrostephanus longispinus* Peters. *Z. vergl. Physiol.* 59:403–412, 1968.

369. MILLOTT, N. Animal photosensitivity, with special reference to eyeless forms. *Endeavour* 16:19–28, 1957.

370. MACKIE, G. O. Pigment effector cells in a cnidarian. *Science* 137:689–690, 1962.

371. SCHWYZER, R., and C. H. LI. A new synthesis of the pentapeptide L-histidyl-L-phenylalanyl-L-arginyl-L-tryptophyl-glycine and its melanocyte-stimulating activity. *Nature* 182:1669–1670, 1958.

372. HOFMANN, K., H. YAJIMA, and E. T. SCHWARTZ. Observations regarding structural prerequisites for melanocyte-expanding activity. *Fed. Proc.* 18:247, 1959.

373. BAGNARA, J. T. Stimulation of melanophores and guanophores by melanophore-stimulating hormone peptides. *Gen. Comp. Endocrinol.* 4:290–294, 1964.

374. SCHNABEL, E., and C. H. LI. The synthesis of L-histidyl-D-phenylalanyl-L-arginyl-L-tryptophyl-glycine and its melanocyte-stimulating activity. *J. Am. Chem. Soc.* 82:4576–4579, 1960.

375. KOIDA, M., K. HANO, and T. ISO. Evaluation of *in vitro* melanocyte-darkening activities of L-histidyl-L-phenylalanyl-L-arginyl-L-tryptophyl-glycine and its nine stereoisomers in *Rana nigromaculata. Jap. J. Pharmacol.* 16:243–249, 1966.

376. SUTHERLAND, E. W., I. ØYE, and R. W. BUTCHER. The action of epinephrine and the role of the adenyl cyclase system in hormone action. *Recent Progr. Hormone Res.* 21:623–642, 1965.

377. SUTHERLAND, E. W., G. A. ROBISON, and R. W. BUTCHER. Some aspects of the biological role of adenosine 3′, 5′-monophosphate (cyclic AMP). *Circulation* 37:279–306, 1968.

378. BITENSKY, M. W., and S. R. BURSTEIN. Effects of cyclic adenosine monophosphate and melanocyte-stimulating hormone on frog skin *in vitro. Nature* 208:1282–1284, 1965.

379. NOVALES, R. R., and W. J. DAVIS. Melanin-dispersing effect of adenosine 3′, 5′-monophosphate on amphibian melanophores. *Endocrinology* 81:283–290, 1967.

380. GOLDMAN, J. M., and M. E. HADLEY. The *beta* adrenergic receptor and cyclic 3′, 5′-adenosine monophosphate: possible roles in the regulation of melanophore responses of the spadefoot toad, *Scaphiopus couchi. Gen. Comp. Endocrinol.* 13:151–163, 1969.

381. ABE, K., R. W. BUTCHER, W. E. NICHOLSON, C. E. BAIRD, R. A. LIDDLE, and G. W. LIDDLE. Adenosine 3′, 5′-monophosphate (cyclic AMP) as the mediator of the actions of melanocyte stimulating hormone (MSH) and norepinephrine on the frog skin. *Endocrinology* 84:362–368, 1969.

382. SCHIMMER, B. P., S. K. UEDA, and G. H. SATO. Site of action of adrenocorticotropic hormone (ACTH) in adrenal cell cultures. *Biochem. Biophys. Res. Commun.* 32:806–810, 1968.

383. PASTAN, I., J. ROTH, and V. MACCHIA. Binding of hormone to tissue: the first step in polypeptide hormone action. *Proc. Nat. Acad. Sci.* 56:1802–1809, 1966.

384. PASTAN, I. and V. MACCHIA. Mechanism of thyroid-stimulating hormone action. *J. Biol. Chem.* 242:5757–5761, 1967.

385. LANGLEY, J. N. On the reaction of cells and of nerve-endings to certain poisons, chiefly as regards the reaction of striated muscles to nicotine and to curare. *J. Physiol.* 33:374–413, 1905.

386. AHLQUIST, R. P. A study of the adrenotropic receptors. *Am. J. Physiol.* 153: 586–600, 1948.

387. GOLDMAN, J. M., and M. E. HADLEY. *In vitro* demonstration of adrenergic receptors controlling melanophore responses of the lizard, *Anolis carolinensis. J. Pharmacol. Exp. Ther.* 166:1–9, 1969.

388. IGA, T. Action of catecholamines on the melanophores in the teleost. *Zool. Mag.* 77:19–26, 1968.

389. DALE, H. H. On some physiological actions of ergot. *J. Physiol.* 34:163–206, 1906.

390. FUJII, R. and R. R. NOVALES. Cellular aspects of the control of physiological color changes in fishes. *Amer. Zool.* 9:453–463, 1969.

391. GOLDMAN, J. M., and M. E. HADLEY. Direct agonistic and antagonistic effects of ergotamine on vertebrate melanophores. *Arch. Int. Pharmacodyn. Ther.* 183:239–246, 1970.

392. HADLEY, M. E., and J. M. GOLDMAN. Adrenergic receptors and geographic variation in *Rana pipiens* chromatophore responses. *Am. J. Physiol.* 219:72–77, 1970.

393. MOORE, J. A. Hybridization between *Rana palustris* and different geographical forms of *Rana pipiens. Proc. Nat. Acad. Sci.* 32:209–212, 1946.

394. GRAHAM, J. D. P. The response to catecholamines of the melanophores of *Xenopus laevis. J. Physiol.* 158:5p–6p, 1961.

395. ROBISON, G. A., R. W. BUTCHER, and E. W. SUTHERLAND. Adenyl cyclase as an adrenergic receptor. *Ann. N.Y. Acad. Sci.* 139:703–723, 1969.

396. TURTLE, J. R., and D. M. KIPNIS. An adrenergic receptor mechanism for the control of cyclic 3′, 5′-adenosine monophosphate synthesis in tissues. *Biochem. Biophys. Res. Commun.* 28:797–802, 1967.

397. MARINETTI, G. W., T. K. RAY, and V. TOMASI. Glucagon and epinephrine stimulation of adenyl cyclase in isolated rat liver plasma membranes. *Biochem. Biophys. Res. Commun.* 36:185–193, 1969.

398. GOLDMAN, J. M., and M. E. HADLEY. Evidence for separate receptors for melanophore-stimulating hormone and catecholamine regulation of cyclic AMP in the control of melanophore responses. *Brit. J. Pharmacol.* 39:160–166, 1969.

399. MARSLAND, D. The mechanism of pigment displacement in unicellular chromatophores. *Biol. Bull.* 87:252–261, 1944.

400. MARSLAND, D., and D. MEISNER. Effects of D_2O on the mechanism of pigment dispersal in the melanocytes of *Fundulus heteroclitus*: a pressure-temperature analysis. *J. Cell. Physiol.,* 70:209–216, 1967.

401. MALAWISTA, S. E. On the action of colchicine. The melanocyte model. *J. Exp. Med.* 122:361–384, 1965.

402. ALEXANDER, N. J., and W. H. FAHRENBACH. The dermal chromatophores of *Anolis carolinensis* (Reptilia, Iguanidae). *Amer. J. Anat.* 126:41–56, 1969.

403. WISE, G. E. Ultrastructure of amphibian melanophores after light-dark adaptation and hormonal treatment. *J. Ultrastr. Res.* 27:472–485, 1969.

404. JANDE, S. S. Fine structure of tadpole melanophores. *Anat. Rec.* 154:533–544, 1966.

405. LERNER, A. B., and Y. TAKAHASHI. Hormonal control of melanin pigmentation. *Recent Progr. Hormone Res.* 12:303–320, 1956.

406. HOROWITZ, S. B. The effects of sulfhydryl inhibitors and thiol compounds on pigment aggregation and dispersion in the melanophores of *Anolis carolinensis*. *Exp. Cell Res.* 13:400–402, 1957.

407. POTTER, R. B., and M. E. HADLEY. Comparative effects of sulfhydryl inhibitors on melanosome movements within vertebrate melanophores. *Experientia* 26:536–538, 1970.

408. JUNQUEIRA, L., and K. R. PORTER. Mechanisms for pigment migration in melanophores and erythrophores. *J. Cell Biol.* 43:62a, 1969.

409. WEBB, J. L. *Enzyme and metabolic inhibitors.* Vol. 2, pp. 332–338. Academic Press, New York, 1966.

410. SCHWARTZ, I. L., H. RASMUSSEN, M. A. SCHOESSLER, L. SILVER, and C. T. O. FONG. Relation of chemical attachment to physiological action of vasopressin. *Proc. Nat. Acad. Sci.* 46:1288–1298, 1960.

411. NOVALES, R. R. The effects of osmotic pressure and sodium concentration on the response of melanophores to intermedin. *Physiol. Zool.* 32:15–28, 1959.

412. GOLDMAN, J. M., and M. E. HADLEY. Cyclic AMP and adrenergic receptors in melanophore responses to methylxanthines. *Europ. J. Pharmacol.* 12:365–370, 1970.

413. ROTHMAN, S., H. F. KRYSA, and A. M. SMILJANIC. Inhibitory action of human epidermis on melanin formation. *Proc. Soc. Exp. Biol. Med.* 62:208–209, 1946.

414. HALPRIN, K. M., and A. OHKAWARA. Glutathione and human pigmentation. *Arch. Derm.* 94:355–357, 1966.

415. HALPRIN, K. M., and A. OHKAWARA. Human pigmentation: the role of glutathione. *Advan. Biol. Skin* 8:241–251, 1967.

416. CHIAN, L. T. Y., and G. F. WILGRAM. Tyrosinase inhibition; its role in suntanning and in albinism. *Science* 155:198–200, 1966.

417. LERNER, A. B., T. B. FITZPATRICK, E. CALKINS, and W. H. SUMMERSON. Mammalian tyrosinase; the relationship of copper to enzymatic activity. *J. Biol. Chem.* 187:793–802, 1950.

418. LEE, T. H., and M. S. LEE. Studies on MSH-induced melanogenesis: effect of long-term administration of MSH on the melanin content and tyrosinase activity. *Endocrinology* 88:155–164, 1971.

419. NOVALES, R. R. The role of ionic factors in hormone action on the vertebrate melanophore. *Am. Zool.* 2:337–352, 1962.

420. SPAETH, R. A. Evidence proving the melanophore to be a disguised type of smooth muscle cell. *J. Exp. Zool.* 20:193–215, 1916.

421. VESELY, D. L., and M. E. HADLEY. (Personal observations, *Rana pipiens*.)

422. NOVALES, R. R., B. J. NOVALES, S. H. ZINNER, and J. A. STONER. The effects of sodium, chloride, and calcium concentration on the response of melanophores to melanocyte-stimulating hormone (MSH). *Gen. Comp. Endocrinol.* 2:286–295, 1962.

423. VESELY, D. L., and M. E. HADLEY. Calcium requirement for melanophore-stimulating hormone action on melanophores. *Science* 173:923–925, 1971.

424. DIKSTEIN, S., C. P. WELLER, and F. G. SULMAN. Effect of calcium ions on melanophore dispersal. *Nature* 200:1106, 1963.

425. DIKSTEIN, S., and F. G. SULMAN. Mechanism of melanophore dispersion. *Biochem. Pharmacol.* 13:819–826, 1964.

426. NOVALES, R. R., and B. J. NOVALES. Factors influencing the response of isolated dogfish skin melanophores to melanocyte-stimulating hormone. *Biol. Bull.* 131:470–478, 1966.

427. FINGERMAN, M. Perspectives in crustacean physiology. Scientia 105(699–700): 1–23. In *Pigmentation: its genesis and control*, V. Riley (Ed.), Appleton-Century-Crofts, New York, 1972.

428. FINGERMAN, M., M. MIYAWAKI, and C. OGURO. Effects of osmotic pressure and cations on the response of the melanophores in the fiddler crab, *Uca pugnax*, to the melanin-dispersing principle from the sinus gland. *Gen. Comp. Endocrinol.* 3:496–504, 1963.

429. FINGERMAN, M., and P. M. CONNELL. The role of cations in the actions of the hormones controlling the red chromatophores of the prawn, *Palaemonetes vulgaris*. *Gen. Comp. Endocrinol.* 10:392–398, 1968.

430. FREEMAN, A. R., P. M. CONNELL, and M. FINGERMAN. An electrophysiological study of the red chromatophore of the prawn *Palaemonetes*: observations on the action of red pigment-concentrating hormone. *Comp. Biochem. Physiol.* 26:1015–1029, 1968.

431. KINOSITA, H. Studies on the mechanism of pigment migration within fish melanophores with special reference to their electric potentials. *Anat. Zool. Jap.* 26:115–127, 1953.

432. HADLEY, M. E., and J. M. GOLDMAN. Effects of cyclic 3′, 5′-AMP and other adenine nucleotides on the melanophores of the lizard, *Anolis carolinensis*. *Brit. J. Pharmacol.* 37:650–658, 1969.

433. FALK, G., and R. W. GERARD. Effect of microinjection of salts and ATP on the membrane potential and mechanical response of muscle. *J. Cell. Comp. Physiol.* 43:393–403, 1954.

434. RICHARDS, C. M., D. T. TARTOF, and G. W. NACE. A melanoid variant in *Rana pipiens*. *Copeia* (1969) pp. 850–852, 1969.

435. BAGNARA, J. T., and S. NEIDLEMAN. Effect of chromatophorotropic hormone on pigments of anuran skin. *Proc. Soc. Exp. Biol. Med.* 97:671–673, 1958.

436. PROTA, G. M. D'AGOSTINO, and G. MISURACA. Isolation and characterization of hallochrome, a red pigment from the sea worm, *Halla parthenopeia*. *Experientia* 27:15–16, 1971.

437. MORIMOTO, I., M. I. N. SHAIKH, R. H. THOMSON, and D. G. WILLIAMSON. The structure of arenicochromine, a trihydroxymethoxybenzpyrenequinone from *Arenicola marina. Chem. Commun.* pp. 550–551, 1970.

438. HAMILTON, III, W. J., and F. HEPPNER. Radiant solar energy and the function of black homeotherm pigmentation: an hypothesis. *Science* 155:196–197, 1967.

439. OHMART, R. D., and R. C. LASIEWSKI. Roadrunners: energy conservation by hypothermia and absorption of sunlight. *Science* 172:67–69, 1971.

440. ROHRLICK, S. T., and K. R. PORTER. Fine structural observations to the production of color by the iridophores of a lizard, *Anolis carolinensis. J. Cell Biol.* 53:38-52, 1972.

441. McGUIRE, J., and G. MOELLMANN. Cytocholasin B: Effects on microfilaments and movement of melanin granules within melanocytes. *Science* 175:642–644, 1972.

442. MALAWISTA, S. E. Cytocholasin B reversibly inhibits melanin granule movement in melanocytes. *Nature* 234:354–355, 1971.

443. CELIS, M. E., S. TALEISNIK, and R. WALTER. Regulation of formation and proposed structure of the factor inhibiting the release of melanocyte-stimulating hormone. *Proc. Nat. Acad. Sci.* 68:1428–1433, 1971.

444. NAIR, R. M. G., A. J. KASTIN, and A. V. SCHALLY. Isolation and structure of hypothalamic MSH release-inhibiting hormone. *Biochem. Biophys. Res. Commun.* 43:1376–1381, 1971.

445. BOWER, S. A., M. E. HADLEY, and V. J. HRUBY. Comparative MSH release-inhibiting activity of tocinoic acid (the ring of oxytocin), and L-Pro-L-Leu-Gly-NH$_2$ (the sidechain of oxytocin). *Biochem. Biophys. Res. Commun.* 45:1185–1191, 1971.

446. HRUBY, V. J., C. W. SMITH, SR. A. BOWER, and M. E. HADLEY. MSH release inhibition by ring structures of neurohypophysial hormones. *Science* 176, 1972.

447. CELIS, M. E., and S. TALEISNIK. *In vitro* formation of a MSH-releasing agent by hypothalamic extracts. *Experienta* 27:1481–1482, 1971.

index

A

Acetylcholine:
  action on chromatophores, 110, 113, 144
  effect on adrenal catecholamine release, 114
  effect on MSH release, 89, 113
Adaptive significance of pigmentation, 165, 166
Addison's disease, 79
Adenine, 5, 17
Adenosine triphosphate (ATP), 158, 159
Adrenal corticotropic hormone (ACTH), effects on pigmentation, 80, 134
  melanogenesis in goldfish, 82
  relationship to MSH, 75, 147
Adrenalectomy, 80
Adrenaline (epinephrine) (*See* Catecholamines)
Adrenal (medullary) catecholamines, 114
Adrenal steroids, 79, 82, 114

Adrenergic blocking agents, 137, 164
Adrenergic nerve fibers, 89, 91
Adrenergic receptors, 113, 137
Agalychnis dacnicolor (Mexican tree frog):
  dermal chromatophore unit, 38
  giant melanosomes, 14, 16
Agouti coat color, 42, 49
Albedo, 29, 31
Ambystoma, 151
  maculatum, 23, 98
  mexicanum, 59, 62
  opacum, 102, 98
Ameiurus nebulosus, 83, 106
Amelanotic melanophores, 25
Annelids, 129
Anolis carolinensis, 39, 77, 84, 87, 111, 146
Anthozoans, 163
Arenicochromine, 163
Arenicola marina, 163
Areolar pigmentation, 115
Artemia, 56
Astaxanthin, 55
Atropine, 144

Autonomic nervous system, 109, 113

B

Background adaptation:
  in crustaceans, 123, 126
  role of MSH in fishes, 83
  theories of chromatophore control, 88
Bathygobius, 87
Bihumoral theory of color control, 88
Birds, 21, 34, 41, 46, 56, 85, 109, 117, 120
Birefringence, 18
Black pigment-concentrating hormone (BPCH), 125
Black pigment-dispersing hormone (BPDH), 125
Blinding, effect on pigmentation, 106, 108
Blue coloration of animals, 1, 17, 35, 41
Body blanching reaction, 99
Briston betularia, 129
Bufo, 31
  arenarum, 91
  bufo, 91
  punctatus, 23
Bursicon, 128
Butterflies, 52, 129

C

Caffeine, 132, 152
Calcium ion, 151, 154

Calliphora, 128
Canthaxanthin, 55
Capsanthin, 55
Carassius auratus (Moor goldfish), 15
Carausius morosus, 129
Carotenes, 55
Carotenoids:
  dietary requirement for coloration, 24, 56
  xanthophores and erythrophores, 22, 23
Castration effects on pigmentation, 115
Catecholamines:
  adrenergic receptors, 140, 164
  inhibition of MSH release, 89, 91
  neuronal control of chromatophores, 110
Caucasoid pigmentation, 9, 149
Cephalopod chromatophore organs:
  iridophores, 5, 18, 21, 45
  morphology, 27
  physiological control, 121
Chameleons, 84, 109
  Chameleo jacksoni, 84
  Chameleo pumilus, 111
Cholinergic mechanisms, 56, 110, 113
Chromatoblasts, 62, 67
Chromatophore control, 74–131
Chromatophores:
  "contraction" and "expansion," 2
  differentiation, 67
  interaction, 62
  metaplasia, 66
  origin, 60
Chromatophorotropins, 124, 125
Chromogenic activity, 63
Clear cells (*See* Amelanotic melanophores)

Cnidarians, 131
Colchicine, 146
Colias eurytheme, 129
Conger orbignyanus, 91
Corethra, 129
Cortisol (hydrocortisone), 114
Counter-shading, 86
Crangon, 123
Crotalus viridis, 84
Crustaceans, 16, 123–127, 153–156
Cryptic coloration, 129
Cushing's disease, 80
Cuticular coloration, 126, 128
Cyclic AMP:
  ionic considerations, 152, 154
  relationship to adrenergic
    receptors, 141
  second messenger in hormone
    stimulation, 133, 141, 147,
    164
Cyclostomes, 84, 85, 99, 108, 112,
  146
Cysteinyldopa, 49
Cytocrine melanin, 7, 8, 9, 10, 12,
  14, 31, 33, 43, 115

D

Daphnia, 56
Dasyatis sabina, 66
Dermal chromatophore unit, 37,
  141
Dermal melanophores, 6
  adrenergic receptors, 137
  dermal chromatophore unit, 37
  embryonic origin, 59
  general aspects, 6–16
  morphological color change, 31
  pituitary control, 74*ff*.
  response to melatonin, 104

Diadema setosum, 130
Dibenamine, 137
Dietary insufficiency, 50
Dimorphic pigment patterns, 115,
  117
Diurnal rhythms:
  crustaceans, 123
  lampreys, 108
Dopa, 47
Drosophila, 25, 50, 58
Drosopterins, 5, 23, 25, 53

E

Ecdysone, 128
Echinoderms, 50, 130
Elasmobranchs, 85, 94, 108, 111,
  151
Electrophoretic theory, 156
Energy requirements for hormone
  action, 157–159
Epidermal melanin unit, 37, 42
Epidermal melanophores
  (melanocytes):
  adrenergic receptors, 141
  development of, 12
  lack of pineal influence on, 104
  in morphological color change,
    32, 82
  morphology, 8
  pigment, 5
  pigmentary patterns, 65
Epinephrine (adrenaline) (*See*
  Catecholamines)
Epiphysis (*See* Pineal)
Ergotamine, 137
Erythrophores, 22, 34
Estradiol, 115
Eumelanin, 47, 49
Eunicella, 163

Excitement darkening, 113, 140
Excitement pallor, 113, 137
Eye shine, 21
Eyestalk chromatophorotropins,
    123, 126

F

Feathers, 1, 9, 11, 42, 47, 85, 117
Fishes, 11, 14, 22, 25, 56, 73, 82,
    84, 87, 91, 105, 109, 112,
    119, 139, 146, 151, 156,
    158
Flavins, 2, 46, 52
Function of skin pigmentation,
    165, 166
Fundulus heteroclitus, 83, 105, 146

G

Geotria australis, 84, 109
Gillichthys mirabilis, 87
Glial cells, 27
Glutathione, 149
Glycolysis, 157
Golgi, 21
Gonadal hormones, 3, 114–117
Green coloration, 1, 22, 35, 41
Guanine, 2, 4, 17, 31, 52
Guinea pig, 115

H

Hair, 32, 80, 104, 115
Halla parthenopeia, 163

Heavy water (D_2O), 145, 146
Hemidactylus brookii, 84
Hippolysmata (*See* Frontispiece)
Hippolyte, 123
Hormone receptors, 136, 141
Hydrocortisone (*See* Cortisol)
Hyla, 31
    arborea, 151
    arenicolor, 23, 39, 86
    cinerea, 18, 24, 40
    regilla, 79
    sylvatica, 23
Hyperpigmentation, 80, 91, 115,
    149
Hypophysectomy:
    effects on pigmentation, 79, 81,
        83, 87, 108
Hypophysial placode, 69
Hypothalamus:
    inhibition of MSH release, 89,
        113, 164
    lesions, 91, 94
Hypoxanthine, 5, 52

I

Idothea montereyensis, 126
Imidazole group, 52
Indole-5, 6-quinone, 47
Ink sac, 21
Innervation:
    of chromatophores, 109, 122
    of the pars intermedia, 189
Insects, 16, 50, 127
Intermediate lobe, 91, 95
Intermedin, 75, 85, 86, 125, (*See
    also* MSH)
Ionic requirements for MSH
    action, 150, 165
Ionone ring, 55

Iridescence, 1, 17
Iridiphores:
   association with other
        chromatophores, 45
   dermal chromatophore unit, 39
   embryonic origin, 59
   influence of thyroid on, 117
   lack of a pineal influence on,
        104
   morphology, 17–21
   in physiological color change,
        86
   in structural coloration, 35
Iris pigment, 21
Isopods, 126
Isoproterenol, 138

Larval melanophores, 31, 52
Leucophore, 21
Light, direct effect on
        chromatophores, 74, 99,
        130
Lipophores, 22, 56
Lithium ion, 151
Lizards, 77, 78, 84, 95, 108, 136,
        139, 158
Locustana, 129
Locusts, 127, 129
Loligo:
   opalescens, 122
   vulgaris, 122
Lutein, 55
Luteinizing hormone (LH), 117

J

Juvenile coat color, 115

K

Klauberina riversiana, 95
Kosciuscola tristis, 127
Kynurenine, 50

M

Malpighian cells, 42
Melanins, 2, 16, 46
Melanoblasts, 31
Melanogenesis, 47, 82, 115, 149
"Melanoiridophore," 26
Melanophage, 26
Melanophore index (MI), 35,
        101, 113
Melanophores (*See* Dermal
        melanophores *and*
        Epidermal melanophores)
Melanophore stimulating
        hormone (MSH,
        intermedin):
   assays, 36, 77
   control of bright colored
        chromatophores, 85*ff*.
   control of melanophores, 79*ff*.
   mechanisms of action, 132, 147
   regulation of release, 89
   chemical structures, 75

L

Lamellar-reflecting ribbons, 5, 18
Lampetra, 84
   planeri, 108
Langerhans cell, 25

Melanosome (melanin granule):
aggregation (contraction) and
dispersion (expansion) of,
33
development, 14
structure, 9, 14, 16
Melatonin, 88, 99, 144
Methylxanthines, 136, 141
Microtubules, 144, 146
Mollusks, 50, 121
Mongoloids, 9
Mordacia mordax, 109
Morphological color change,
31–33, 70, 81, 97, 119, 128
Moths, 129
MSH releasing factor (MRF), 96,
164
MSH release-inhibiting factor
(MRIF), 96, 164
Mustelus canis, 108
Myxine, 84
glutinosa, 146

N

Negroids, 9, 80, 149
Nereis dumerilii, 129
Nervous control:
of chromatophores, 28, 83, 87,
109
of the pars intermedia, 91, 109,
113
Neural crest, 58, 66
Neurointermediate lobe, 91
Neurosecretory nerve fibers, 89,
94
Neurotransmitter agents, 112
Nevi, 112
Norepinephrine (noradrenaline),
89, 113 (*See also*
Catecholamines)

Notophthalamus viridescens, 22,
44, 59
Nuptial coloration, 3, 22, 117,
166

O

Oedipoda coerulescens, 127
Ommatidia, 50, 126
Ommatins, 50
Ommins, 50
Ommochromes, 5, 16, 46, 50
Origin of pigment cells, 60
Oryzias latipes, 56
Ouabain, 152

P

Palaemonetes, 123, 153, 156
Pandulus jordani, 124
Parasympathetic nervous system,
110, 113, 114
Parietal eye, 101, 107
Parr-smolt transformation, 73,
120
Pars distalis, 77, 95
Pars intermedia, 77, 85, 86, 89,
95, 113, 164
Pars nervosa, 95
Patterns of pigmentation, 9, 12,
22, 42, 43, 59, 62, 65, 69,
70, 73, 80, 111
Pelage pigmentation (*See* Hair)
Periplaneta americana, 128
Peromyscus maniculatus, 80
Phaeomelanins, 5, 47, 49
Phentolamine, 137
Phosphodiesterase, 136, 144, 165

Photometric reflectance method, 36, 77
Photoreceptors, 101
Photosensitivity of chromatophores, 130
Phoxinus, 87, 98, 106
Phrynosoma, 108, 111
Physiological color changes, 22, 28, 32, 33, 43, 81, 86, 127
Pigment cell development, 161
Pigment granule movements, 144
Pineal, 88, 98, 102
Pinealectomy, 102, 104, 109
Pituitary gland, 74, 83, 109
Plethodon cinereus, 23, 26
Pleurodeles, 70
Plumage pigmentation, 3, 32, 117
Poephila gouldiae, 56
Porphyrins, 2
Premelanosome, 12, 14
Primary stage of chromatic control, 29
Progesterone, 115
Prolactin, 83, 87
Pteridines, 2, 5, 25, 46, 51, 54
Pterinosomes, 5, 24, 26, 66
Purines, 5, 17, 21, 32, 46, 51, 54, 161
Pyrazine ring, 52

R

Raia, 111
Raja radiata, 94
Rana:
　esculenta, 104
　pipiens, 7, 8, 12, 31, 37, 44, 61, 64, 77, 99, 102, 134
　nigromaculata, 91
　sylvatica, 68
　temporaria, 89

Red hair, 50
Red pigment-concentrating hormone (RPCH), 125, 153
Red pigment-dispersing hormone (RPDH), 125, 153
Reflecting platelets, 5, 18, 75
Reptiles, 11, 14, 22, 84, 87, 100, 111, 112, 117, 140
Resting stage of melanophores, 157
Retinal melanophores, 127
Rhina squatima, 111
Rhythms of color change, 108, 123, 127
Riboflavin, 54

S

Salmo gairdneri, 110
Sarcophaga, 128
Scaphiopus couchi, 140
Scylliorhinus stellaris, 94
Sea urchin, 130
Secondary stage of chromatic control, 29, 31
Second messenger hypothesis, 133
Sepiapterins, 5, 23, 25, 53
Serotonin, 125
Sheath cells, 27
Sinus gland, 124
Sodium ion requirement for hormone action, 150, 153
Sol-gel transformations, 144, 150
Squalus acanthias, 75
Steroids, 79, 114, 117
Stirnorgan, 100
Stress (*See* Excitement darkening *and* Excitement pallor)
Structural coloration, 1, 17, 34
　diffraction and interference, 34

Sulfhydryl agents (inhibitors), 146, 164
Sympathetic nervous system, 88, 136

T

Tanning agent, 128
Tapetum lucidum, 66
Taricha torosa, 102
Tautogolabrous adspersus, 110
Temperature, effects on coloration, 123, 159
Terminology, 4
Testosterone, 115
Tetrodotoxin, 110, 153, 156
Theophylline, 134, 142, 152
Theories of chromatophore control, 88
Thermoregulation, 127, 129, 166
Thyroidectomy, 119
Thyroxine, 73, 118
Transplantation experiments, 60
Trichosiderins, 49
Triturus, 22
  pyrrhogaster, 53
Tryptophan, 50
Tyndall scattering, 1, 17, 34, 41
Tyrosinase, 14, 47, 149

U

Uca, 123, 161
Uca pugnax, 21, 153
Ultraviolet light, 149, 165

Unihumoral theory of color control, 88
Uric acid, 5, 17, 52

V

Vitiligo, 80, 112

W

White pigment-concentrating hormone (WPCH), 125
White pigment-dispersing hormone (WPDH), 125
W-substance, 88

X

Xanthic goldfish, 66
Xantholeucophore, 45
Xanthophores:
  associations, 45
  dermal chromatophore unit, 39
  embryonic origin, 59
  general aspects, 22–24
  in green coloration, 41
  in physiological color change, 34
  response to intermedin, 75, 86
Xanthophylls, 55
Xenopus laevis, 7, 29, 66, 68, 99, 102, 113
Xiphophorus, 25
X-organ-sinus gland, 124